Automatic Control Systems

Automatic Control Systems

Editor

Bianca Lupei

Automatic Control Systems
Edited by **Bianca Lupei**

Printed in 2017w Te t

ISBN: 978-1-68117-109-8
Library of Congress Control Number: 2015950369

© 2016 by
SCITUS Academics LLC,
616, Corporate Way, Suite 2, 4766,
Valley Cottage, NY 10989

www.scitusacademics.com

Notice

Table of Content

Preface

Automatic control is the application of control theory for regulation of processes without direct human intervention. In the simplest type of an automatic control loop, a controller compares a measured value of a process with a desired set value, and processes the resulting error signal to change some input to the process, in such a way that the process stays at its set point despite disturbances. This closed-loop control is an application of negative feedback to a system.

The book, Automation Control Systems, prepares carefully, the background of each topic. Detail explanations of each step given make it easy for the students to understand the complicated derivations. Most of the theoretical topic are supported with practical illustrations.

It covers emerging technologies in a timely manner for two important subjects automatic control systems. In particular, it allows researchers and students to become familiar with state-of-the-art developments in the field, and it also helps engineers and practitioners enhance their productivity with the latest technology.

CHAPTER 1

Optimal Gravitational Search Algorithm for Automatic Generation Control of interconnected Power Systems

Rabindra Kumar Sahu, ,Sidhartha Panda , Saroj Padhan

Department of Electrical Engineering, Veer Surendra Sai University of Technology (VSSUT), Burla 768018, Odisha, India

ABSTRACT

An attempt is made for the effective application of Gravitational Search Algorithm (GSA) to optimize PI/PIDF controller parameters in Automatic Generation Control (AGC) of interconnected power systems. Initially, comparison of several conventional objective functions reveals that ITAE yields better system performance. Then, the parameters of GSA technique are properly tuned and the GSA control parameters are proposed. The superiority of the proposed approach is demonstrated by comparing the results of some recently published techniques such as Differential Evolution (DE), Bacteria Foraging Optimization Algorithm (BFOA) and Genetic Algorithm (GA). Additionally, sensitivity analysis is carried out that demonstrates the robustness of the optimized controller parameters to wide variations in operating loading condition and time constants of speed governor, turbine, tie-line power. Finally, the proposed approach is extended to a more realistic power system model by considering the physical constraints such as reheat turbine, Generation Rate Constraint (GRC) and Governor Dead Band nonlinearity.

INTRODUCTION

The main objective of a power system utility is to maintain continuous supply of power with an acceptable quality to all the consumers in the system. The system will be in equilibrium, when there is a balance between the power demand and the power generated. There are two basic control mechanisms used to achieve power balance; reactive power balance (acceptable voltage profile) and real power balance (acceptable frequency values). The former is called the Automatic Voltage Regulator (AVR) and the latter is called the Automatic Load Frequency Control (ALFC) or Automatic Generation Control (AGC). For multiarea power systems, which normally consist of interconnected control area, AGC is an important aspect to keep the system frequency and the interconnected area tie-line power as close as possible to the intended values [1]. The mechanical input power to the generators is used to control the system as it is affected by the output electrical power demand and to maintain the power exchange between the areas as planned. AGC monitors the system frequency and tie-line flows, calculates the net change in the generation required according to the change in demand and changes the set position of the generators within the area so as to keep the time average of the ACE (Area Control Error) at a low value. ACE is generally treated as controlled output of AGC. As the ACE is adjusted to zero by the AGC, both frequency and tie-line power errors will become zero [2].

Several control strategies for AGC of power systems have been proposed in order to maintain the system frequency and tie line power flow at their scheduled values during normal and disturbed conditions. In [3], a critical literature review on the AGC of power systems has been presented. It is observed that, considerable research work is going on to propose better AGC systems based on modern control theory [4], neural network [5], fuzzy system theory [6], reinforcement learning [7] and ANFIS approach [8]. But, these advanced approaches are complicated and need familiarity of users to these techniques thus reducing their applicability. Alternatively, a classical Proportional Integral Derivative (PID) controller and its variant remain an engineer's preferred choice due to its structural simplicity, reliability, and the favorable ratio between performances and cost. Additionally, it also offers simplified dynamic modeling, lower user-skill requirements, and minimal development effort, which are major issues of in engineering practice. In recent times, new artificial intelligence-based approaches have been proposed to optimize the PI/PID controller parameters for AGC system. In [9], several classical controllers structures such as Integral (I), Proportional Integral (PI), Integral Derivative (ID), PID and Integral Double Derivative (IDD) have been applied and their performance has been compared for an AGC system. Nanda et al. [10] have demonstrated that Bacterial Foraging Optimization Algorithm (BFOA) optimized controller provides better performance than GA based controllers and conventional controllers for an interconnected power system. In [11], Ali and Abd-Elazim have employed a BFOA to optimize the PI controller parameters and shown its superiority over GA in a two area non-reheat thermal system. A gain scheduling PI controller for an AGC system has been proposed by Gozde and Taplamacioglu [12] for a two area thermal power system with governor dead-band nonlinearity where the authors have employed a Craziness based Particle Swarm Optimization (CPSO) with different objective functions to minimize the settling times and standard error criteria. Shabani et. al [13] employed an Imperialist Competitive Algorithm (ICA) to optimize the PID controller parameters in a multiarea multiunit power system. In [14], a modified objective function using Integral of Time multiplied by Absolute value of Error (ITAE), damping ratio of dominant eigenvalues and settling time is proposed where the PI controller parameters are optimized employed Differential Evolution (DE) algorithm and the results are compared with BFOA and GA optimized ITAE based PI controller to show its superiority.

It obvious from literature survey that, the performance of the power system not only depends on the artificial techniques employed but also depends on the controller structure and chosen objective function. Hence, proposing and implementing new high performance heuristic optimization algorithms to real world problems are always welcome. Gravitational Search Algorithm (GSA) is a newly developed heuristic optimization method based on the law of gravity and mass interactions [15]. It has been reported in the literature that GSA is more efficient in terms of CPU time and offers higher precision with more consistent results [16]. However, studied on choosing the controller parameters of GSA has not been reported in the literature. In a PID controller, the derivative mode improves stability of the system and increases speed of the controller response but it produces unreasonable size control inputs to the plant. Also, any noise in the control input signal will result in large plant input signals which often lead to complications in practical applications. The practical solution to these problems is to put a first filter on the derivative term and tune its pole so that the chattering due to the noise does not occur since it attenuates high frequency noise. Surprisingly, in spite of these advantages, Proportional Integral Derivative with derivative Filter (PIDF) controller structures are not attempted for the AGC problems. Having known all this, an attempt has been made in the present paper for the optimal design of GSA based PI/PIDF controller for AGC in a multiarea interconnected power system.

The aim of the present work is as follows:

(i) to study the effect of objective function of the system performance
(ii) to tune the control parameters of GSA
(iii) to demonstrate the advantages of GSA over other techniques such as DE, BFOA and GA which are recently reported in the literature for the similar problem
(iv) to show advantages of using a modified controller structure and objective function to further increase the performance of the power system
(v) to study the effect of the physical constraints such as Generation Rate Constraints and governor dead band on the system performance.

SYSTEM MODELING

The system under investigation consists of two area interconnected power system of non-reheat thermal plant as shown in Fig. 1. Each area has a rating of 2000 MW with a nominal load of 1000 MW. The system that is widely used in the literature is for the design and analysis of automatic load frequency control of interconnected areas. In Fig. 1, B_1 and B_2 are the frequency bias parameters; ACE_1 and ACE_2 are area control errors; u_1 and u_2 are the control outputs form the controller; R_1 and R_2 are the governor speed regulation parameters in pu Hz; T_{G1} and T_{G2} are the speed governor time constants in s; ΔP_{V1} and ΔP_{V2} are the change in governor valve positions (pu); ΔP_{G1} and ΔP_{G2} are the governor output command (pu); T_{T1} and T_{T2} are the turbine time constant in s; ΔP_{T1} and ΔP_{T2} are the change in turbine output powers; ΔP_{D1} and ΔP_{D2} are the load demand changes; ΔP_{Tie} is the incremental change in tie line power (p.u); K_{PS1} and K_{PS2} are the power system gains; T_{PS1} and T_{PS2} are the power system time constant in s; T_{12} is the synchronizing coefficient and Δf_1 and Δf_2 are the system frequency deviations in Hz,. The relevant parameters are given in Appendix A.

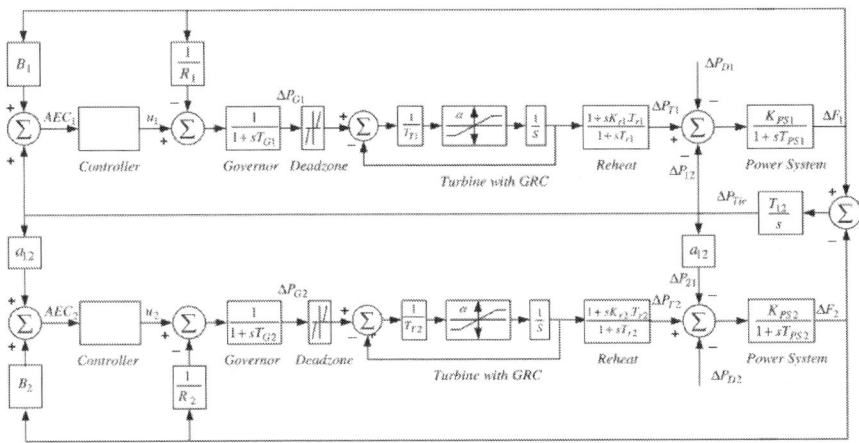

Figure 1. Transfer function model of two-area non-reheat thermal system.

Each area of the power system consists of speed governing system, turbine and generator as shown in Fig. 1. Each area has three inputs and two outputs. The inputs are the controller input ΔP_{ref} (denoted as u_1 and u_2), load disturbances (denoted as ΔP_{D1} and ΔP_{D2}), and tie-line power error ΔP_{Tie}. The outputs are the generator frequency deviations (denoted as ΔF_1 and ΔF_2) and Area Control Error (ACE) given by [2].

$$ACE = B\Delta F + \Delta PT\ ie \tag{1}$$

where B is the frequency bias parameter.

To simplicity the frequency-domain analyses, transfer functions are used to model each component of the area. Turbine is represented by the transfer function [2]:

$$G_T(s) = \frac{\Delta P_T(s)}{\Delta P_V(s)} = \frac{1}{1 + sT_T} \tag{2}$$

From [2], the transfer function of a governor is as follows:

$$G_G(s) = \frac{\Delta P_V(s)}{\Delta P_G(s)} = \frac{1}{1 + sT_G} \tag{3}$$

The speed governing system has two inputs ΔP_{ref} and ΔF with one output $\Delta P_G(s)$ given by [2]:

$$\Delta P_G(s) = \Delta P_{ref}(s) - \frac{1}{R}\Delta F(s) \tag{4}$$

The generator and load is represented by the transfer function [2]:

$$G_P(s) = \frac{K_P}{1 + sT_P} \tag{5}$$

where $K_P = 1/D$ and $T_P = 2H/fD$.

The generator load system has two inputs $\Delta P_T(s)$ and $\Delta P_D(s)$ with one output $\Delta F(s)$ given by [2]:

$$\Delta F(s) = G_P(s)[\Delta P_T(s) - \Delta P_D(s)] \tag{6}$$

OVERVIEW OF GRAVITATIONAL SEARCH ALGORITHM

Gravitational Search Algorithm (GSA) is one of the newest heuristic algorithms inspired by the Newtonian laws of gravity and motion [15]. In GSA, agents are considered as objects and their performance is measured by their masses. All these objects attract each other by the force of gravity and this force causes a global movement of all objects toward the objects with a heavier mass. Hence masses co-operate using a direct form of communication through gravitational force. The heavy masses that correspond to good solution move more slowly than lighter ones, and this guarantees the exploitation step of the algorithm.

In GSA, each mass (agent) has four specifications: position, inertial mass, active gravitational mass and passive gravitational mass. The position of the mass corresponds to a solution of the problem and its gravitational and inertia masses are determined using a fitness function. In other words each mass presents a solution and the algorithm is navigated by properly adjusting the gravitational and inertia masses. By lapse of time it is expected that masses be attracted by the heavier mass. This mass will present an optimum solution in the search space. The GSA could be considered as an isolated system of masses. It is like a small artificial world of masses obeying the Newtonian laws of gravitation and motion. Masses obey the following laws [15] and [16].

Law of gravity
Each particle attracts every other particle and the gravitational force between the two particle is directly proportional to the product of their masses and inversely proportional to the distance between them R. It has been reported in the literature that R provides better results than R^2 in all experiment cases [15].

Law of motion
The current velocity of any mass is equal the sum of the fraction of its previous velocity and the variation in the velocity. Variation in the velocity or acceleration of any mass is equal to the force acted on the system divided by mass of inertia.

For a system with 'n' agent (masses), the ith position of an agent X_i is defined by:

$$X_i = (x_i^1, \ldots, x_i^d, \ldots x_i^n) \quad \text{for} \quad i = 1, 2, \ldots n \tag{7}$$

where, x_i^d is the represents the position of ith agent in the dth dimension.

At a specific time 't', the force acting on mass 'i' from mass 'j' is defined as follows:

$$F_{ij}^d(t) = G(t) \frac{M_{pi}(t) * M_{aj}(t)}{R_{ij}(t) + \epsilon} \left(x_j^d(t) - x_i^d(t) \right) \tag{8}$$

where, M_{aj} is the active gravitational mass related to agent j, M_{pi} is the passive gravitational mass related to agent i, $G(t)$ is the gravitational constant at time t, ϵ is small constant, and $R_{ij}(t)$ is the Euclidian distance between two agents i and j given by:

$$R_{ij}(t) = X_{i(t)}, X_{j(t)2} \tag{9}$$

The stochastic characteristic in GSA algorithm is incorporated by assuming that the total forces that act on agent 'i' in a dimension 'd' be a randomly weight sum of dth components of the forces exerted from other agents as follows:

$$F_i^d(t) = \sum_{j=1, j \neq i}^{n} rand_j F_{ij}^d(t) \tag{10}$$

where $rand_j$ is a random number in the interval [0, 1]

The acceleration of the agent 'i' at the time t and in the direction dth, is given by the law of the motion as:

$$a_i^d(t) = \frac{F_i^d(t)}{M_{ii}(t)}$$

(11)

where $M_{ii}(t)$ is the inertia mass of ith agent.

The velocity of an agent is updated depending on the current velocity and acceleration. The velocity and position are updated as follows:

$$v_i^d(t+1) = rand_i * v_i^d(t) + a_i^d(t)$$

(12)

$$x_i^d(t+1) = x_i^d(t) + v_i^d(t+1)$$

(13)

where $rand_i$ is a uniform random variable in the interval (0, 1). The random number is used to give a randomized characteristic to the search process.

The gravitational constant G is initialized at the beginning. To control the search accuracy it is reduced with time and expressed as function of the initial value (G_0) and time t as:

$$G(t) = G_0 e^{(-\alpha t/T)}$$

(14)

where a is a constant and T is the number of iteration.

The masses (gravitational and inertia) are evaluated by the fitness function. Efficient agents are characterized by heavier masses. Assuming the equal gravitational and inertia mass, the values of masse are calculated using the map of fitness. The gravitational and inertial masses are updated as follows:

$$M_{ai} = M_{pi} = M_{ii} = M_{i}, \quad i = 1, 2, \ldots n.$$

(15)

$$m_i(t) = \frac{fit_i(t) - worst(t)}{best(t) - worst(t)} \tag{16}$$

$$M_i(t) = \frac{m_i(t)}{\sum_{j=1}^{N} m_j(t)} \tag{17}$$

where $fit_i(t)$ represents the fitness value of the agent 'i' at time t and $best(t)$ is defined for a minimization problem as:

$$Best(t) = \min_{j \in \{1...n\}} fit_j(t) \tag{18}$$

$$Worst(t) = \max_{j \in \{1...n\}} fit_j(t) \tag{19}$$

To achieve a good compromise between exploration and exploitation, the number of agents is reduced with lapse of Eq. (10) and therefore a set of agents with bigger mass are used for applying their force to the other.

The performance of GSA is improved by controlling exploration and exploitation. To avoid trapping in a local optimum GSA must use the exploration at beginning. By lapse of iterations, exploration must fade out and exploitation must fade in. In GSA only the Kbest(which is a function of time, with the initial value K_0 at the beginning and decreasing with time) agents attract the others. At the beginning, all agents apply the force, and as time passes Kbest is decreased linearly and at the end there is just one agent applying force to the others. Therefore, Eq. (10) is modified as follows:

$$F_i^d(t) = \sum_{j \in kbest, j \neq i} rand_j F_{ij}^d(t) \tag{20}$$

where Kbest is the set of first K agents with the best fitness value and biggest mass k.

The different steps of the GSA are the followings:

  I.   Identify the search space of parameters to be searched.
  II.   Initialize the variables.

III.   Evaluate the fitness of each agent.
IV.   Update $G(t)$, $best(t)$, $worst(t)$ and $M_i(t)$ for $i = 1, 2, \ldots, n$.
V.   Calculate the total force in various directions.
VI.   Calculate the acceleration and velocity.
VII.   Update the position of the agents.
VIII.   Repeat steps (iii) to (vii) until the stop criteria is reached.
IX.   End.

GSA is characterized as a simple concept which is easy to implement and computationally efficient. In order to improve exploration and exploitation capabilities, GSA has a flexible and balanced mechanism. More precise search is achieved by assuming a higher inertia mass which causes a slower motion of agents in the search space. Faster convergence is obtained by considering a higher gravitational mass which causes a higher attraction of agents. GSA is a memory-less algorithm but works powerfully like the other memory based algorithms. The nature inspired population based techniques have proved themselves to be effective solutions to optimization problems control parameters and objective function is involved in these optimization techniques, and appropriate selection of these is a key point for success. It has been reported that, GSA tends to find the global optimum faster than other algorithms and has a higher convergence rate for uni-modal high-dimensional functions. The performance of GSA for multi-modal functions is comparable with other algorithms [15].

THE PROPOSED APPROACH

Controller structure
The Proportional Integral Derivative Controller (PID) is the most popular feedback controller used in the process industries. It is a robust, easily understood controller that can provide excellent control performance despite the varied dynamic characteristics of process plant. As the name suggests, the PID algorithm consists of three basic modes, the proportional mode, the integral and the derivative modes. A proportional controller has the effect of reducing the rise time, but never eliminates the steady-state error. An integral control has the effect of eliminating the steady-state error, but it may make the transient response worse. A derivative

control has the effect of increasing the stability of the system, reducing the overshoot, and improving the transient response. Proportional integral (PI) controllers are the most often type used today in industry. A control without derivative (D) mode is used when: fast response of the system is not required, large disturbances and noises are present during operation of the process and there are large transport delays in the system. PID controllers are used when stability and fast response are required. Derivative mode improves stability of the system and enables increase in proportional gain and decrease in integral gain which in turn increases speed of the controller response. However, when the input signal has sharp corners, the derivative term will produce unreasonable size control inputs to the plant. Also, any noise in the control input signal will result in large plant input signals. These reasons often lead to complications in practical applications. The practical solution to the these problems is to put a first filter on the derivative term and tune its pole so that the chattering due to the noise does not occur since it attenuates high frequency noise. In view of the above a filter is used for the derivative term in the present paper.

In the present paper, identical controllers have been considered for the two areas as the two areas are identical. The structure of PID controller with derivative filter is shown inFig. 2 where K_P, K_I and K_D are the proportional, integral and derivative gains respectively, and N is the derivative filter coefficient. When used as PI controller, the derivative path along with the filter is removed from Fig. 2. The error inputs to the controllers are the respective area control errors (ACE) given by:

$$e_1(t) = ACE_1 = B_1 \Delta F_1 + \Delta PTie \qquad (21)$$

$$e_2(t) = ACE_2 = B_2 \Delta F_2 - \Delta PTie \qquad (22)$$

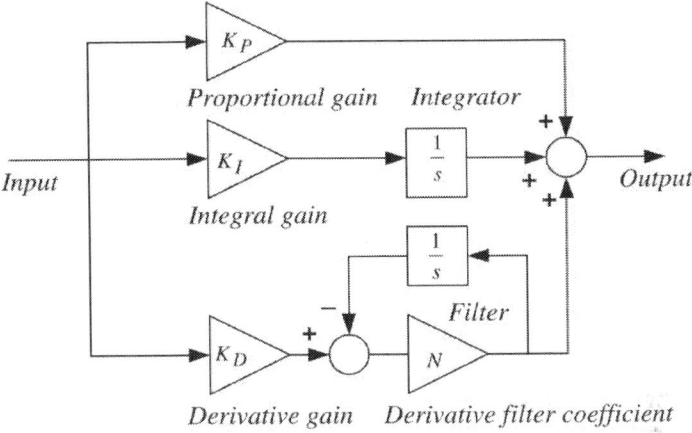

Figure 2. Structure of PID controller with derivative filter.

The control inputs of the power system u_1 and u_2 are the outputs of the controllers. The transfer function of the controller is given by:

$$TF_{PID} = \left[K_P + K_I \left(\frac{1}{s} \right) + K_D \left(\frac{Ns}{s + N} \right) \right]$$

(23)

Objective function

While designing a controller, the objective function is first defined based on the desired specifications and constraints. The design of objective function to tune controller parameters is generally based on a performance index that considers the entire closed loop response. Some of the realistic control specifications for Automatic Generation Control (AGC) are as follows [2]:

(i) The frequency error should return to zero following a load change.

(ii) The integral of frequency error should be minimum.

(iii) The control loop must be characterized by a sufficient degree of stability.

(iv) Under normal operating conditions, each area should carry its own load and the power exchange between control areas following a load perturbation should maintained at its prescheduled value as quickly as possible.

To determining the optimum values of controller parameters conventional objective functions are considered at the first instance. Time-domain techniques based on objective functions can be classified into two groups: (a) Criteria based on a few points in the response (b) Criteria based on the entire response, or integral criteria. The integral criteria are generally accepted as a good measure for system performance. An advantage of using the integral gain is that it can be easily extended to a multi-loop system. The commonly used integral based error criteria are as follows: Integral of Squared Error (ISE), Integral of Absolute Error (IAE), Integral of Time multiplied Squared Error (ITSE) and Integral of Time multiply by Absolute Error (ITAE). These integral based objective functions for the present problem are expressed as given below:

$$J_1 = ISE = \int_0^{t_{sim}} (|\Delta F_1| + |\Delta F_2| + |\Delta P_{Tie}|)^2 \cdot dt \tag{24}$$

$$J_2 = IAE = \int_0^{t_{sim}} (|\Delta F_1| + |\Delta F_2| + |\Delta P_{Tie}|) \cdot dt \tag{25}$$

$$J_3 = ITSE = \int_0^{t_{sim}} (|\Delta F_1| + |\Delta F_2| + |\Delta P_{Tie}|)^2 \cdot t \cdot dt \tag{26}$$

$$J_4 = ITAE = \int_0^{t_{sim}} (|\Delta F_1| + |\Delta F_2| + |\Delta P_{Tie}|) \cdot t \cdot dt \tag{27}$$

In the above equations, ΔF_1 and ΔF_2 are the system frequency deviations; ΔP_{Tie} is the incremental change in tie line power; t_{sim} is the time range of simulation.

The problem constraints are the PI/PIDF controller parameter bounds. Therefore, the design problem can be formulated as the following optimization problem.

$$\text{Minimize} \quad J \tag{28}$$

Subject to

For PI controller : $K_{Pmin} \leq K_P \leq K_{Pmax}, \quad K_{Imin} \leq K_I \leq K_{Imax}$ (29)

For PIDF controller : $K_{Pmin} \leq K_P \leq K_{Pmax},$

$$K_{Imin} \leq K_I \leq K_{Imax}, \quad K_{Dmin} \leq K_D \leq K_{Dmax} \tag{30}$$

where J is the objective function $(J_1, J_2, J_3 \text{ and } J_4)$ and K_{PIDmin} and K_{PIDmax}, are the minimum and maximum value of the PI/PID control parameters. As reported in the literature [10], [11], [12], [13], [14] and [17], the minimum and maximum values of PID controller parameters are chosen as -2.0 and 2.0 respectively. The range for filter coefficient N is selected as 1 and 100 [17].

RESULTS AND DISCUSSIONS

Application of GSA
At the first instance physical constraints such as reheat turbine, Generation Rate Constraint and governor dead band are neglected. In the absence of above physical constraints, the studied power system becomes similar to that used in references[11] and [14]. The model of the system under study is developed in MATLAB/SIMULINK environment and GSA program is written (in .mfile). The developed model is simulated in a separate program (by .mfile using initial population/controller parameters) considering a 10% step load change in area 1. The objective function is calculated in the .mfile and used in the optimization algorithm. At the first instance, the following parameters are chosen for the application of GSA: population size NP = 30; maximum iteration = 500; gravitational constants G_0 = 30 and a = 10; K_0 = total number of agents and decreases linearly to 1 with time [18]. Optimization is terminated by the prespecified number of generations. The flowchart of proposed optimization is shown in Fig. 3. Simulations were conducted on an Intel, core 2 Duo CPU of 2.4 GHz and 2 GB MB RAM computer in the MATLAB 7.10.0.499 (R2010a) environment. The optimization was repeated 50 times and

the best final solution among the 50 runs is chosen as final controller parameters. The best final solutions obtained in the 50 runs for each objective functions are shown inTable 1. To investigate the effect of objective function on the dynamic performance of the system, settling times (2% of final value) and peak overshoots in frequency and tie-line power deviations along with minimum damping ratios are also provided in Table 1. It can be seen from Table 1 that best system performance is obtained with maximum value of damping ratio and minimum values of settling times and peak overshoots in frequency and tie-line power deviations when ITAE is used as objective function.

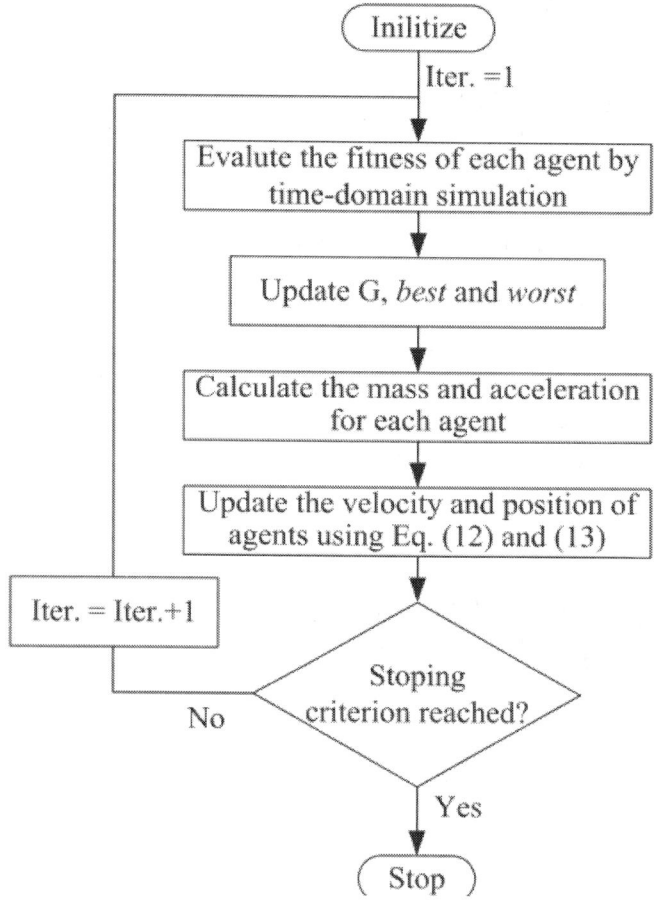

Figure 3. Flow chart of proposed GSA optimization approach.

Table 1. Tuned controller parameters, settling time, peak overshoot and minimum damping ratio for each objective function.

Objective function	Controller parameters		T_S (s)			Peak overshoot			ζ
	Proportional gain (K_P)	Integral gain (K_I)	ΔF_1	ΔF_2	ΔP_{Tie}	ΔF_1	ΔF_2	ΔP_{tie}	
J_1:ISE	−0.0120	0.8641	21.30	21.30	15.10	0.0594	0.0877	0.0152	0.0626
J_2:IAE	−0.0117	0.7668	18.00	18.00	12.90	0.0450	0.0710	0.0119	0.0733
J_3:ITSE	−0.1228	0.7849	16.20	16.20	11.10	0.0499	0.0838	0.0127	0.0882
J_4:ITAE	−0.1701	0.6492	**12.00**	**11.90**	**8.90**	**0.0390**	**0.0590**	**0.0083**	**0.1144**

Bold signifies the best results.

GSA parameters tuning

The success of GSA is heavily dependent on setting of control parameters namely; constant a, initial gravitational constant G_0, population size NP and number of iteration T. While applying GSA these control parameters should be carefully chosen for the successful implementation of the algorithm. A series of experiments were conducted to properly tune the GSA control parameters in order to optimize the PI parameters employing ITAE objective function. Table 2 shows the GSA outcomes as a result of varying its control parameters. To quantify the results, 50 independent runs were executed for each parameter variation. It is clear from results shown in Table 2 that the best settings for constant a, gravitational constant G_0, population size NP and number of iteration T are $a = 20$, $G_0 = 100$, NP $= 20$ and $T = 100$ respectively. Note that increasing the population size NP beyond 20 and iterations T beyond 100 will improve the average, maximum and standard deviation values slightly (with same minimum value) at the expense of increasing the computation time significantly.

Table 2. Study of tuning GSA parameters.

Parameter	Min	Ave	Max	St. Dev	Other parameters
$x = 10$	0.6659897	0.8980626	1.0123918	0.1265678	$NP = 20, T = 50, G_0 = 100$
$x = 15$	0.6659897	0.7841139	1.0001651	0.1267459	
$x = 20$	**0.6659897**	**0.7291873**	**0.9182768**	**0.0802699**	
$x = 25$	0.6749712	0.9000713	0.9754091	0.1228549	
$x = 30$	0.6870195	0.9164638	1.0123918	0.1375834	
$G_0 = 30$	0.6659945	0.8046012	1.0123918	0.1184529	$NP = 20, T = 50, x = 20$
$G_0 = 70$	0.6659988	0.8512434	1.0225763	0.1282661	
$G_0 = 100$	**0.6659897**	**0.7291873**	**0.9182768**	**0.0802699**	
$G_0 = 130$	0.6680531	0.7790509	1.0225763	0.1055190	
$G_0 = 150$	0.6678854	0.7991217	1.0035452	0.1988937	
$NP = 10$	0.6881325	0.8231840	1.0315944	0.1988937	$T = 50, x = 20, G_0 = 100$
$NP = 15$	0.6706229	0.8019416	0.9426240	0.1250767	
$NP = 20$	**0.6659897**	**0.7291873**	**0.9182768**	**0.0802699**	
$NP = 25$	0.6659897	0.7280129	0.9146560	0.0702121	
$NP = 30$	0.6659897	0.7279858	0.9135081	0.0679424	
$T = 30$	0.6650960	0.8781403	0.9742163	0.1040270	$NP = 20, G_0 = 100, x = 20$
$T = 50$	0.6659897	0.7291873	0.9182768	0.0802699	
$T = 100$	**0.6659897**	**0.6774539**	**0.8739539**	**0.0571679**	
$T = 200$	0.6659897	0.6764037	0.8739539	0.0570019	

Bold signifies the best results.

Modified objective function

Once the GSA control parameters are set, modifications in the objective function and controller structure are considered to further improve the performance of the power system. The modified objective function J_5 tries to minimize the ITAE error, maximizes the minimum damping ratios of dominant eigenvalues and minimizes the settling times of Δf_1, Δf_2 and ΔP_{Tie} as given by Eq. (31) below:

$$J_5 = \int_0^{t_{sim}} \omega_1 \cdot \int_0^{t_{sim}} (|\Delta F_1| + |\Delta F_2| + |\Delta P_{Tie}|) \cdot t \cdot dt$$

$$+ \omega_2 \cdot \frac{1}{\min \left(\sum_{i=1}^{n} (1 - \zeta_i) \right)} + \omega_3(ST) \tag{31}$$

where, ΔF_1 and ΔF_2 are the system frequency deviations; ΔP_{Tie} is the incremental change in tie line power; t_{sim} is the time range of simulation; ζ_i is the damping ratio and n is the total number of the dominant eigenvalues; ST is the sum of the settling times of frequency and tie line power deviations respectively; ω_1 to ω_3 are weighting factors. Inclusion of appropriate weighting factors to the right hand individual terms helps to make each term competitive during the optimization process. Wrong choice of the weighting factors leads to incompatible numerical values of each term involved in the definition of fitness function which gives misleading result. The weights are so chosen that numerical value of all the terms in the right hand side of Eq. (31) lie in the same range. Repetitive trial runs of the optimizing algorithms are executed with both PI and PIDF controller to select the weights. To make each term competitive during the optimization process the following weights are chosen: PI controller: $\omega_1 = 1.0$, $\omega_2 = 0.3$ and $\omega_3 = 0.08$ and PIDF controller: $\omega_1 = 1.0$, $\omega_2 = 0.05$ and $\omega_3 = 0.02$. A 10% step load change in area 1 is considered at $t = 0$ s and the optimum PI and PIDF controller parameters with ITAE and modified objective function are obtained employing tuned GSA parameters as given in Table 3 and Table 4. The ITAE values, minimum damping ratios and settling times (2% of final value) for above controllers are also provided in Table 4. For comparison, the results of some recently published technique/controller/objective function for the same power system are also given in the Table 3 and Table 4.

Table 3. Tuned controller parameter and error with ITAE objective function.

Technique	Tuned controller parameter				
	K_P	K_I	K_D	N	ITAE
GA:PI [11]	−0.2346	0.2662	–	–	2.747
BFOA:PI [11]	−0.4207	0.2795	–	–	1.827
DE:PI [14]	−0.2146	0.4335	–	–	0.991
GSA:PI	−0.1880	0.6179	–	–	0.666
GSA:PIDF	1.1884	1.9589	0.3456	54.33	**0.117**

Bold signifies the best results.

Table 4. Settling time, minimum damping ratio and error with modified objective function.

Technique	Tuned controller parameter				Settling time T_s (s)			ζ	ITAE
	K_P	K_I	K_D	N	ΔF_1	ΔF_2	ΔP_{tie}		
DE:PI [14]	−0.4233	0.2879	–	–	5.38	6.95	6.21	0.2361	1.6766
GSA:PI	−0.4383	0.3349	–	–	5.17	6.81	4.59	0.2374	1.3096
GSA:PIDF	1.4011	1.9981	0.7102	93.2760	**1.92**	**3.19**	**2.86**	**0.4470**	**0.1362**

Bold signifies the best results.

Analysis of results

It is clear from Table 3 that with PI structured controller and ITAE objective function (J_4), minimum ITAE value is obtained with GSA (ITAE = 0.6659) compared to ITAE values with GA (ITAE = 2.74), BFOA (ITAE = 1.827) and DE (ITAE = 0.9911) techniques. So it can be concluded that for the similar controller structure (PI) and same objective function (ITAE) GSA outperforms GA, BFOA and DE techniques. It is also evident from Table 3that minimum ITAE value (ITAE = 0.1174) is obtained with a PIDF controller and therefore the performance of PIDF controller is superior to that of PI controller. When ITAE is used as objective function, the system performance in terms of minimum damping ratio and settling times of ΔF_1, ΔF_2 and ΔP_{Tie} with GSA are inferior to GA, BFOA and DE. However, when the modified objective function (J_5) given by Eq. (31) is used better performance is obtained in all respects. The minimum damping ratio ($\zeta = 0.2374$), ITAE value (ITAE = 1.3096) and settling times (5.17, 6.81 and 4.59 s for ΔF_1, ΔF_2 and ΔP_{Tie} respectively) are better compared to those with DE, BFOA and GA technique as shown in Table 4. The best system performance is obtained with GSA optimized PIDF controller optimized using the modified objective function as evident form Table 4. For better visualization of the improvements with the proposed approach, the above results are presented graphically in Fig. 4.

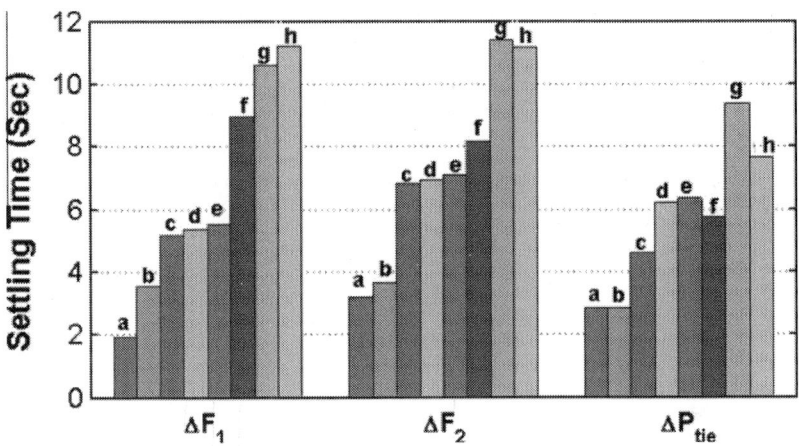

Figure 4. Comparison of settling time (a) GSA PIDF:J₅; (b) GSA PIDF:J₄; (c) GSA PI:J₅; (d) DE PI:J₅; (e) BFOA PI:J₄; (f) DE PI:J₄; (g) GA PI:J₄; (h) GSA PI:J₄.

To study the dynamic performance of the proposed controllers optimized employing tuned GSA using modified objective function (J_5), a step increase in demand of 10% is applied at $t = 0$ s in area-1 and the system dynamic responses are shown in Figure 5,Figure 6 and Figure 7. For comparison, the simulation results with GA and BFOA optimized PI controller using ITAE objective function [11] and DE optimized PI controller using modified objective function [14] for the same power system are also shown inFigure 5, Figure 6 and Figure 7. Critical analysis of the dynamic responses clearly reveals that significant improvement is observed with PIDF controller optimized employing GSA using modified objective function (J_5) compared to GA, BFOA and DE PI.

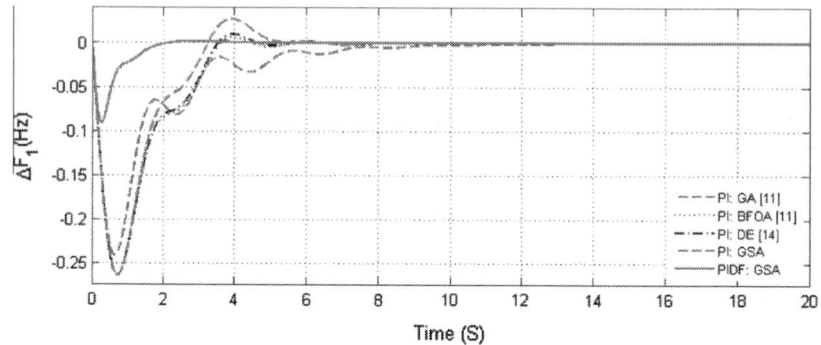

Figure 5. Change in frequency of area-1 for 10% change in area-1.

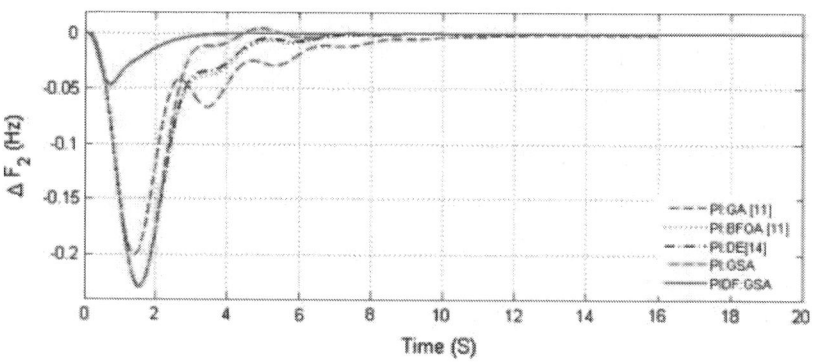

Figure 6. Change in frequency of area-2 for 10% change in area-1.

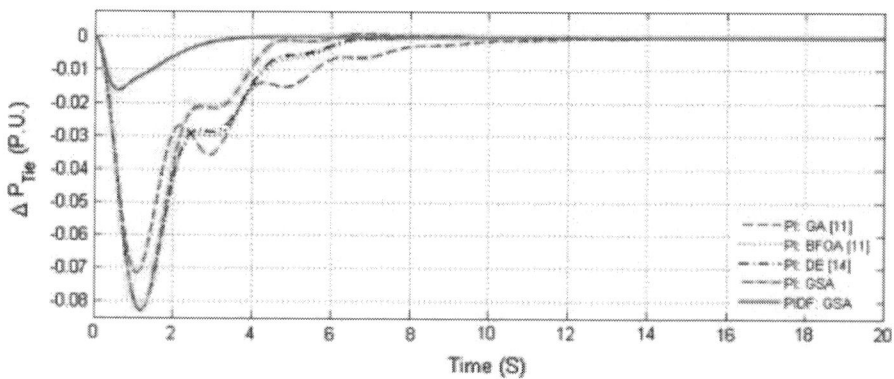

Figure 7. Change in tie line power for 10% change in area-1.

The performance of the proposed controllers is further investigated for simultaneous load disturbance at both areas. A simultaneous step increase in demand of 10% in area-1 and 30% in area-2 is considered at $t = 0.0$ s and the system responses are shown in Figure 8,Figure 9 and Figure 10 from which it is evident that the designed controllers are robust and perform satisfactorily when the location of the disturbance changes.

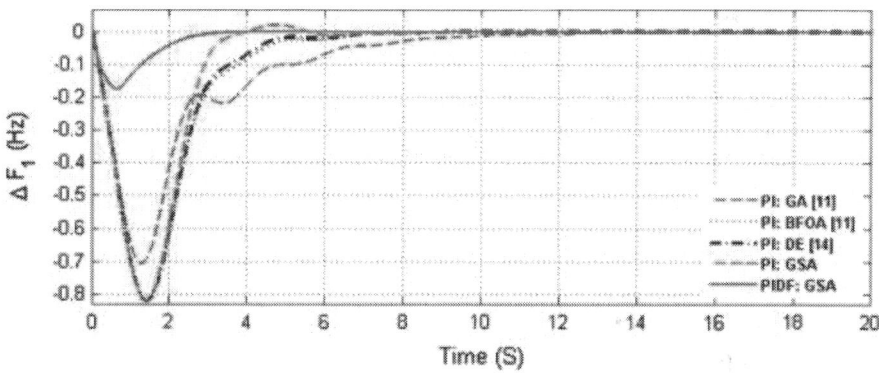

Figure 8. Change in frequency of area-1 for 10% change in area-1 and 30% change in area-2.

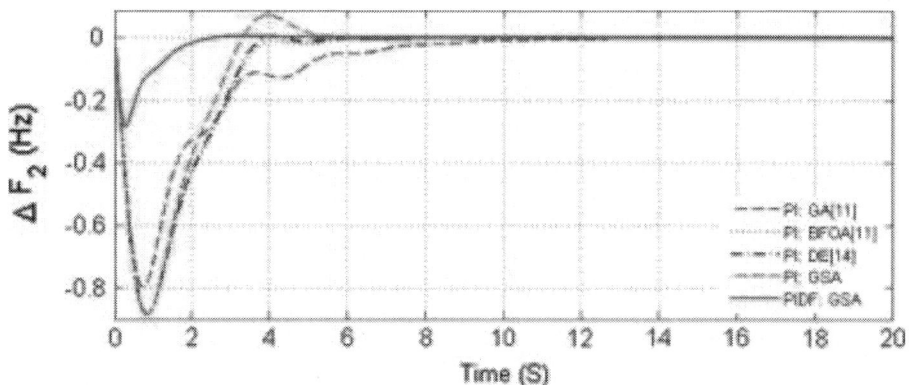

Figure 9. Change in frequency of area-2 for 10% change in area-1 and 30% change in area-2.

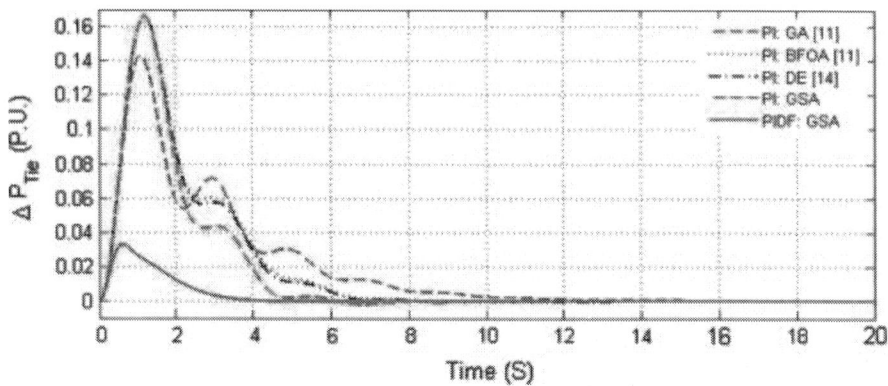

Figure 10. Change in tie line power for 10% change in area-1 and 30% change in area-2.

Sensitivity analysis

Sensitivity analysis is carried out to study the robustness the system to wide changes in the operating conditions and system parameters [9] and [10]. Taking one at a time, the operating load condition and time constants of speed governor, turbine, tie-line power are changed from their nominal values (given in Appendix A) in the range of +50% to −50% in steps of 25%. PIDF controller optimized employing GSA using modified objective function J_5 is considered due to its superior performance.

The optimum values of controller parameters, at changed loading conditions and changed system parameters (for a step increase in demand of 10% at $t = 0$ s in area-1) are provided in Table 5. The corresponding performance indexes (ITAE values, settling times and minimum damping ratios) with the above varied system conditions are given inTable 5. Critical examination of Table 5 clearly reveals that the performance indexes are more or less same. The frequency deviation response of area 1 with above varied conditions is shown in Figure 11, Figure 12, Figure 13 and Figure 14. It can be observed from Figure 11, Figure 12, Figure 13 and Figure 14 that the effect of the variation in operating loading conditions and system time constants on the system responses is negligible. So it can be concluded that, the proposed control strategy provides a robust control and the controller parameters obtained at the nominal loading with nominal parameters, need not be reset for wide changes in the system loading or system parameters.

Table 5. Sensitivity analysis of system without physical constraints.

Parameter variation	% Change	Tuned controller parameter				Settling time, T_s (s)			ζ	ITAE
		K_P	K_I	K_D	N	ΔF_1	ΔF_2	ΔP_{tie}		
Nominal	0	1.4011	1.9981	0.7102	93.2760	1.92	3.19	2.86	0.4470	0.136
Loading condition	+50	1.4103	1.9945	0.7257	89.0014	1.96	3.21	2.88	0.4501	0.138
	+25	1.4081	1.9983	0.7255	92.7390	1.95	3.20	2.87	0.4515	0.137
	−25	1.4067	1.9909	0.7137	86.4434	1.94	3.20	2.87	0.4424	0.137
	−50	1.4178	1.9901	0.7052	93.9838	1.93	3.22	2.88	0.4409	0.136
T_G	+50	1.5001	1.9923	0.7746	95.0011	1.99	3.29	2.92	0.2906	0.137
	+25	1.4386	1.9942	0.7212	93.7709	1.89	3.23	2.88	0.3518	0.134
	−25	1.3293	1.9991	0.6013	94.9823	1.83	3.14	2.79	0.5067	0.131
	−50	1.3112	1.9986	0.6021	90.4946	1.87	3.11	2.79	0.5949	0.131
T_T	+50	1.5792	1.9338	0.8224	91.9358	1.82	3.37	2.99	0.3368	0.134
	+25	1.4975	1.9972	0.7406	94.9587	1.80	3.24	2.88	0.3728	0.132
	−25	1.3002	1.9992	0.6004	95.0012	1.93	3.16	2.81	0.5032	0.136
	−50	1.3009	1.9911	0.6009	94.8310	2.09	3.25	2.86	0.5112	0.148
T_{12}	+50	1.4165	1.9980	0.7496	94.9982	2.19	2.98	2.71	0.3107	0.138
	+25	1.4127	1.9870	0.7006	95.0012	2.10	3.09	2.78	0.3608	0.135
	−25	1.4145	1.9920	0.7002	94.9467	3.57	3.48	3.05	0.5417	0.137
	−50	1.2517	1.9989	0.6025	94.0867	4.10	3.83	3.24	0.6683	0.126

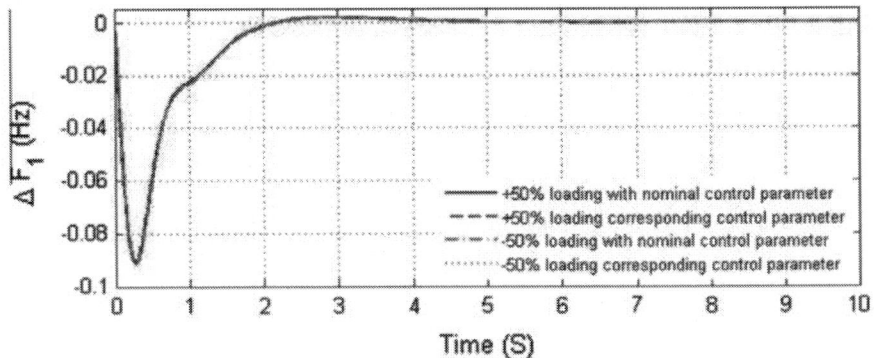

Figure 11. Frequency deviation of area-1 for change in nominal load.

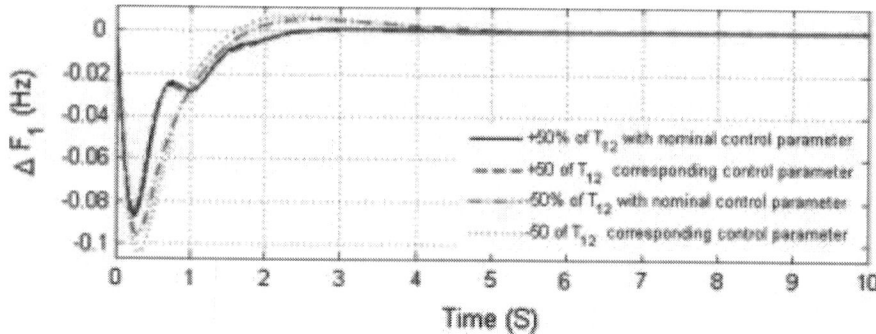

Figure 12. Frequency deviation of area-1 for change in T_{12}.

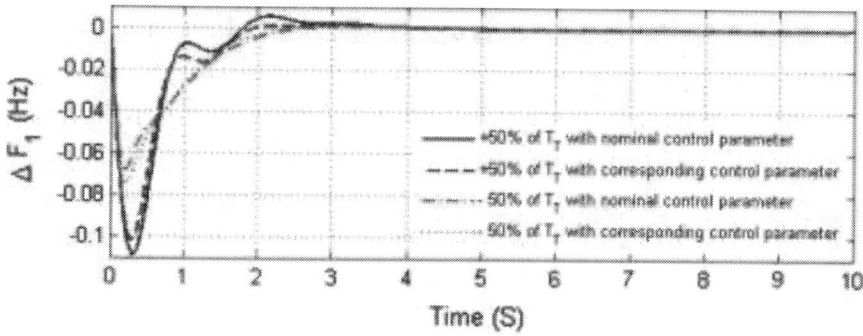

Figure 13. Frequency deviation of area-1 for change in T_T.

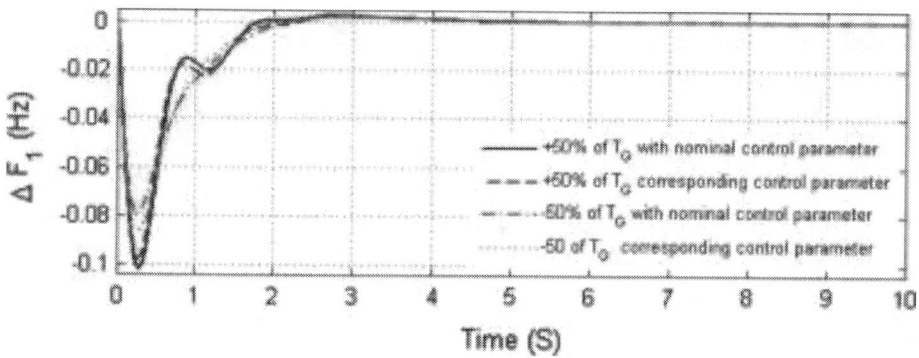

Figure 14. Frequency deviation of area-1 for change in T_G.

Inclusion of physical constraints

To get an accurate insight into the AGC problem, it is necessary to include the important inherent requirement and the basic physical constraints and include them model. The major constraints that affect the power system performance are reheat turbine, Generation Rate Constraint (GRC), and governor dead band (GBD) nonlinearity [19]. In view of the above, the study is further extended to a more realistic power system by considering the effect of reheat turbine, GRC, and GBD. As most of the thermal plants are of reheat type, a reheat turbine is also considered in the proposed realistic power system model. In a power system having steam plants, power generation can change only at a specified maximum rate. The generation rate for non-reheat thermal units is usually higher than the generation rate for reheat units. The reheat units have a generation rate about of 3–10% pu MW/min [20]. The speed governor dead band has a great effect on the dynamic performance of electric energy system. GBD is defined as the total amount of a continued speed change within which there is no change in valve position. The effect of the GBD is to increase the apparent steady-state speed regulation. The speed-governor dead band has makes the system oscillatory. A describing function approach is used to include the GBD nonlinearity. The maximum value of dead band for governors of large steam turbines is typically specified as 0.06% (0.036 Hz) [19]. In view of the above, a GRC of 3%/ min and GBD of 0.036 Hz are considered in the present work.

To investigate the importance of considering the physical constraints, two cases (Case A and Case B) are considered. In Case A, no constraint is considered in the model and in Case B, GBD, GRC and reheat turbine are considered. It is observed that the system becomes unstable when the optimum parameters which were obtained for Case A are applied to the Case B. Hence, the PIDF controller parameters are retuned for Case B employing GSA using modified objective function (J_5) for a 10% step load increase in area-1 at $t = 0$ s. The optimum values of controller parameters are as follows:

$$K_P = 0.8589, \quad K_I = 0.0791, \quad K_D = 1.992, N = 44.5678.$$

A step increase in demand of 10% is applied at $t = 0$ s in area-1 and the system dynamic responses is shown in Figure 15, Figure 16 and Figure 17. It is evident from Figure 15,Figure 16 and Figure 17 that the system is stable with retuned controller parameters but the dynamics of the power system is affected with increased over shoot, performance errors and settling times. Finally, sensitivity analysis is done to study the robustness the system to wide changes in the operating conditions and system parameters as before. The various performance indexes (ITAE values, settling times and minimum damping ratios) under normal and parameter variation cases are given in Table 6. It can be noticed from Table 6 that when physical constraints are introduced, the variations in performance index are more prominent. So it can be concluded that in the presence of GBD, GRC and reheat turbine, the system becomes highly non-linear (even for small load perturbation) and hence performance of the designed controller is degraded. To complete the analysis, a 10% step load increase in area-1 at $t = 0$ s is considered and the frequency deviation response of area-1 for the above varied conditions are shown in Figure 18,Figure 19, Figure 20 and Figure 21. From Table 6 and Figure 18, Figure 19, Figure 20 and Figure 21 it can be concluded that once the controller parameters are tuned under varied conditions, the performance of the proposed controllers that are satisfactory is more or less the same under varied conditions.

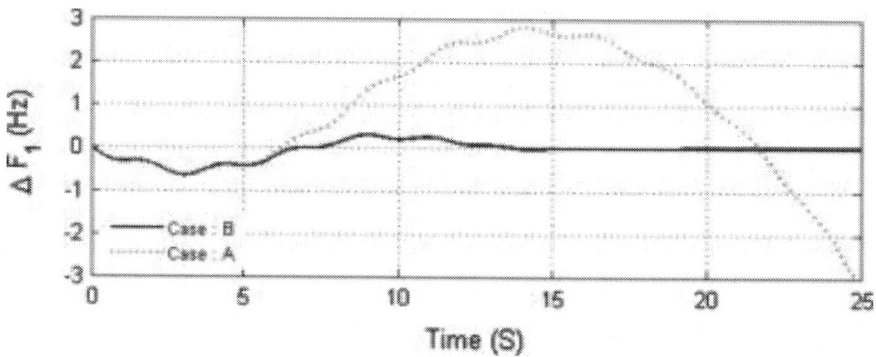

Figure 15. Change in frequency of area-1 for 10% change in area-1.

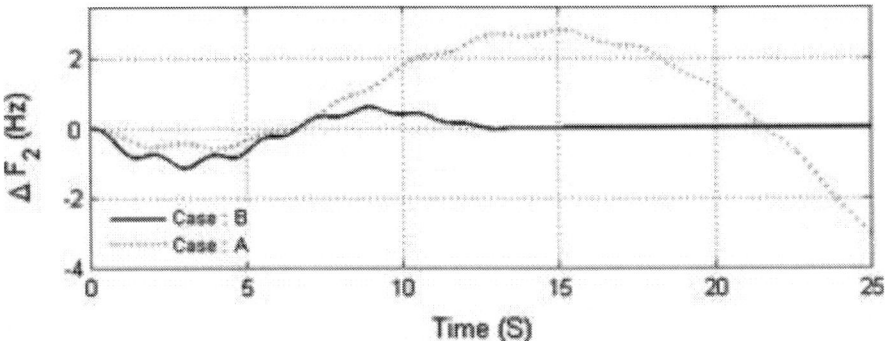

Figure 16. Change in frequency of area-2 for 10% change in area-1.

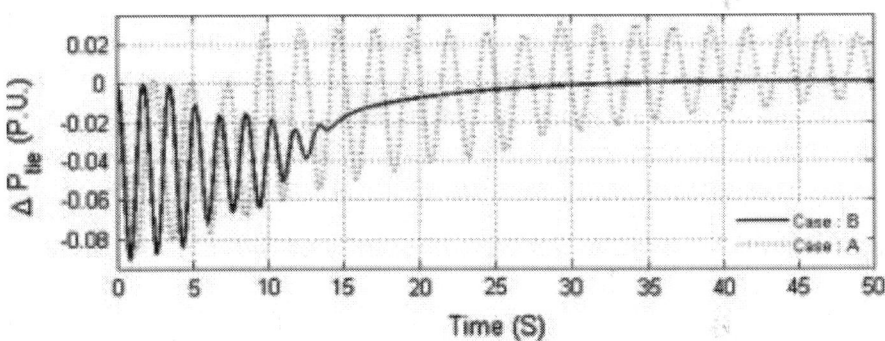

Figure 17. Change in tie line power for 10% change in area-1.

Table 6. Sensitivity analysis of system with physical constraints.

Parameter variation	% Change	Tuned controller parameter				Settling time, T_s (s)			ζ	ITAE
		K_P	K_I	K_D	N	ΔF_1	ΔF_2	ΔP_{tie}		
Nominal	0	0.8589	0.0791	1.9920	44.5678	18.21	19.50	37.44	0.4400	40.4612
Loading condition	+50	0.8630	0.0753	1.9057	43.3623	14.34	13.96	42.37	0.4492	31.4246
	+25	0.8397	0.0758	1.8949	49.1008	16.83	18.98	41.29	0.4670	35.6197
	−25	0.7508	0.0831	1.7503	43.2105	20.56	21.60	34.95	0.4743	49.8424
	−50	0.7500	0.0765	1.9995	48.4272	20.62	21.64	37.78	0.4511	50.4765
T_G	+50	0.8000	0.0800	1.6614	48.8137	19.73	20.51	36.12	0.3802	49.2855
	+25	0.8000	0.0800	1.9168	44.4899	19.57	20.52	36.57	0.3926	43.8609
	−25	0.8811	0.0841	1.8552	44.2036	18.51	19.79	35.09	0.5445	40.4841
	−50	0.9560	0.0855	1.8971	44.4794	17.13	18.43	34.84	0.6684	38.1144
T_T	+50	0.7757	0.0787	1.8821	42.9658	18.03	19.42	37.09	0.4944	40.1139
	+25	0.7908	0.0720	1.9924	47.5757	15.17	19.06	43.01	0.4840	38.3354
	−25	0.7036	0.0749	1.8137	46.2931	16.25	19.15	39.76	0.4297	38.3024
	−50	0.7003	0.0761	1.7980	49.4480	15.85	18.85	38.84	0.3897	38.2674
T_{12}	+50	0.7000	0.0770	1.9987	48.9204	21.54	22.22	38.97	0.3620	47.2833
	+25	0.7000	0.0737	1.9575	49.9986	19.19	20.51	40.96	0.4186	41.6126
	−25	0.7000	0.0684	1.8434	48.6640	17.04	15.48	43.34	0.5116	33.0340
	−50	0.6026	0.0806	1.5000	42.7331	16.29	22.51	30.90	0.5901	35.3342

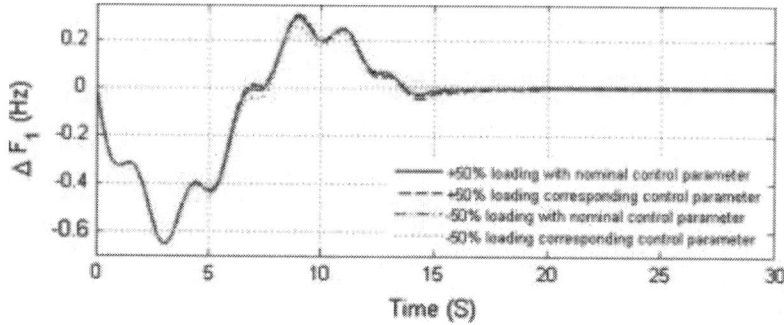

Figure 18. Frequency deviation of area-1 for 10% change in area-1 with physical constraints.

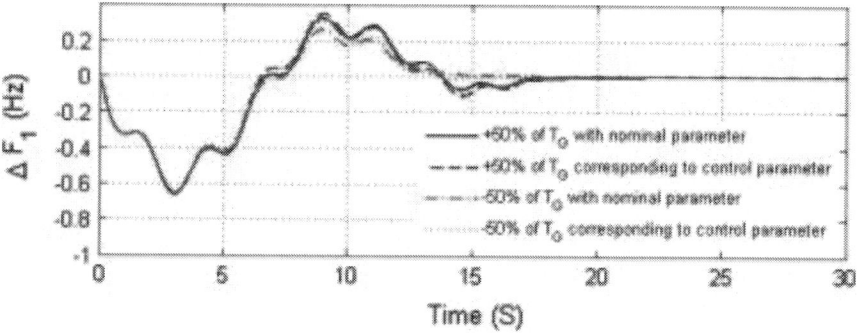

Figure 19. Frequency deviation of area-1 for 10% change in area-1 with physical constraints.

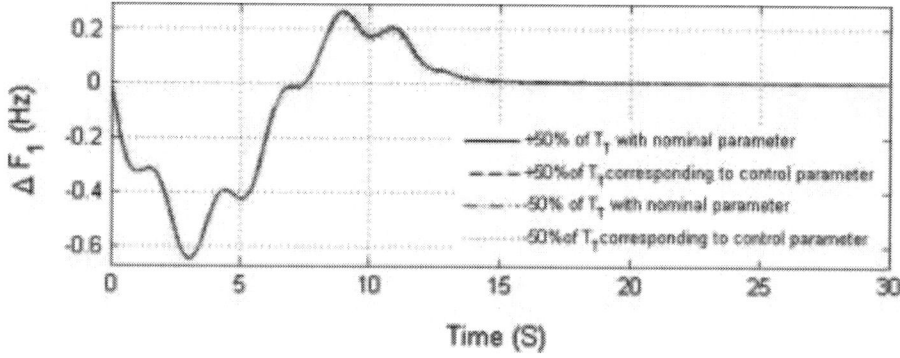

Figure 20. Frequency deviation of area-1 for 10% change in area-1 with physical constraints.

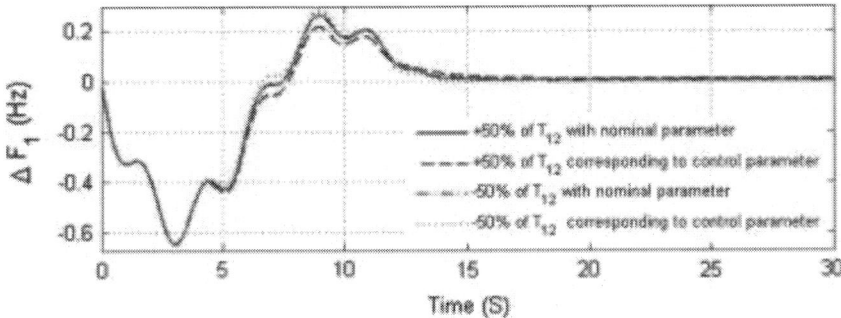

Figure 21. Frequency deviation of area-1 for 10% change in area-1 with physical constraints.

CONCLUSION

An attempt has been made for the first time to apply a powerful computational intelligence technique like GSA to optimize PI and PIDF controller parameters for AGC of a multiarea interconnected power system. Firstly, the system without any physical constraint is optimized using conventional objective functions. It is observed the performance of the power system is better in terms of minimum damping ratio, settling times and peak overshoots in frequency and tie-line power deviations when ITAE objective function is used compared to IAE, ISTE and ISE objective functions. Then, the parameters of GSA technique are properly tuned and the recommended GSA parameters are found to be: $a = 20$, $G_0 = 100$, $NP = 20$ and $T = 100$ respectively. Further, a modified objective function is employed and the parameters of PI and PIDF controller are optimized by tuned GSA. The superiority of the proposed approach is demonstrated by comparing the results with Differential Evolution (DE), Bacteria Foraging Optimization Algorithm (BFOA) and Genetic Algorithm (GA) techniques. Sensitivity analysis reveals that the optimum PIDF controller tuned at the nominal and varied conditions are quite robust and performs satisfactorily under wide changes in system loading conditions or in system parameters. Finally, the proposed approach is extended to a more realistic power system model by considering the physical constraints such as reheat turbine, GRC and governor dead band nonlinearity. It is observed that the when physical constraints are introduced, the variations in performance index are more prominent as evident from the sensitivity analysis.

APPENDIX A

Nominal parameters of the system investigated are: P_R = 2000 MW (rating), P_L = 1000 MW (nominal loading); f = 60 Hz, B_1, B_2 = 0.045 pu MW/Hz; $R_1 = R_2 = 2.4$ Hz/pu; $T_{G1} = T_{G2} = 0.08$ s; $T_{T1} = T_{T2}$ = 0.3 s; $K_{PS1} = K_{PS2} = 120$ Hz/puMW; $T_{PS1} = T_{PS2} = 20$ s; T_{12} = 0.545 pu; $a_{12} = -1$, $K_{r1} = K_{r2} = 0.5$, $T_{r1} = T_{r2} = 10$.

REFERENCES

1. Kundur P. Power system stability and control; 8th reprint. New Delhi: Tata McGraw-Hill; 2009.
2. Elgerd OI. Electric energy systems theory. An introduction. New Delhi: Tata McGraw-Hill; 1983.
3. Ibraheem P, Kothari DP. Recent philosophies of automatic generation control strategies in power systems. IEEE Trans Power Syst 2005;20(1):346–57.
4. Shoults RR, Jativa Ibarra JA. Multi area adaptive LFC developed for a comprehensive AGC simulation. IEEE Trans Power Syst 1993;8(2):541–7.
5. Chaturvedi DK, Satsangi PS, Kalra PK. Load frequency control: a generalized neural network approach. Electr Power Energy Syst 1999;21(6):405–15.
6. Ghosal SP. Optimization of PID gains by particle swarm optimization in fuzzy based automatic generation control. Electr Power Energy Syst 2004;72:203–12.
7. Ahamed TPI, Rao PSN, Sastry PS. A reinforcement learning approach to automatic generation control. Electr Power Energy Syst 2002;63:9–26.
8. Khuntia SR, Panda S. Simulation study for automatic generation control of a multi-area power system by ANFIS approach. Appl Soft Comput 2012;12:333–41.
9. Saikia LC, Nanda J, Mishra S. Performance comparison of several classical controllers in AGC for multi-area interconnected thermal system. Electr Power Energy Syst 2011;33:394–401.

10. Nanda J, Mishra S, Saikia LC. Maiden application of bacterial foraging based optimization technique in multiarea automatic generation control. IEEE Trans Power Syst 2009;24(2):602–9.
11. Ali ES, Abd-Elazim SM. Bacteria foraging optimization algorithm based load frequency controller for interconnected power system. Electr Power Energy Syst 2011;33:633–8.
12. Gozde H, Taplamacioglu MC. Automatic generation control application with craziness based particle swarm optimization in a thermal power system. Electr Power Energy Syst 2011;33:8–16.
13. Shabani H, Vahidi B, Ebrahimpour M. A robust PID controller based on imperialist competitive algorithm for load-frequency control of power systems. ISA Trans 2013;52(1):88–95.
14. Rout UK, Sahu RK, Panda S. Design and analysis of differential evolution algorithm based automatic generation control for interconnected power system. Ain Shams Eng J 2013;4(3):409–21.
15. Rashedi E, Nezamabadi-pour H, SaryazdiJ S. GSA: a gravitational search algorithm. Inf Sci 2009;179:2232–48.
16. Rashedi E, Nezamabadi-pour H, SaryazdiJ S. Filter modeling using gravitational search algorithm. Eng Appl Artif Intell 2011;24:117–22.
17. Tan W. Unified tuning of PID load frequency controller for power systems via IMC. IEEE Trans Power Syst 2010;25(1):341–50.
18. Li Chaoshun, Zhou Jianzhong. Parameters identification of hydraulic turbine governing system using improved gravitational search algorithm. Energy Convers Manage 2011;52:374–81.
19. Golpîra H, Bevrani H, Golpîra H. Application of GA optimization for automatic generation control design in an interconnected power system. Energy Convers Manage 2011;52:2247–55.
20. Sudha KR, Raju YB, Sekhar AC. Fuzzy C-means clustering for robust decentralized load frequency control of interconnected power system with generation rate constraint. Electr Power Energy Syst 2012;37:58–66.

CITATION

Rabindra Kumar Sahu, Sidhartha Panda, Saroj Padhan, Optimal gravitational search algorithm for automatic generation control of interconnected power systems, Ain Shams Engineering Journal, Volume 5, Issue 3, September 2014, Pages 721-733, ISSN 2090-4479, http://dx.doi.org/10.1016/j.asej.2014.02.004.

CHAPTER 2

Differential Evolution Algorithm Based Automatic Generation Control for Interconnected Power Systems with Non-Linearity

Banaja Mohanty[1], Sidhartha Panda[2], P.K. Hota[1]

[1] Department of Electrical Engineering, Veer Surendra Sai University of Technology (VSSUT), Burla 768018, Odisha, India
[2] Department of Electrical and Electronics Engineering, Veer Surendra Sai University of Technology (VSSUT), Burla 768018, Odisha, India

ABSTRACT

This paper presents the design and performance analysis of Differential Evolution (DE) algorithm based Proportional–Integral (PI) and Proportional–Integral–Derivative (PID) controllers for Automatic Generation Control (AGC) of an interconnected power system. Initially, a two area thermal system with governor dead-band nonlinearity is considered for the design and analysis purpose. In the proposed approach, the design problem is formulated as an optimization problem control and DE is employed to search for optimal controller parameters. Three different objective functions are used for the design purpose. The superiority of the proposed approach has been shown by comparing the results with a recently published Craziness based Particle Swarm Optimization (CPSO) technique for the same interconnected power system. It is noticed that, the dynamic performance of DE optimized PI controller is better than CPSO optimized PI controllers. Additionally, controller parameters are tuned at different loading conditions so that an adaptive gain scheduling control strategy can be employed. The study is further extended to a more realistic network of two-area six unit system with different power generating units such as thermal, hydro, wind and diesel generating units considering boiler dynamics for thermal plants, Generation Rate Constraint (GRC) and Governor Dead Band (GDB) non-linearity.

INTRODUCTION

An interconnected power system is made up of several areas and for the stable operation of power systems; both constant frequency and constant tie-line power exchange should be provided. In each area, an Automatic Generation Controller (AGC) monitors the system frequency and tie-line flows, computes the net change in the generation required (generally referred to as area control error – ACE) and changes the set position of the generators within the area so as to keep the time average of the ACE at a low value [1]. Therefore ACE, which is defined as a linear combination of power net-interchange and frequency deviations, is generally taken as the controlled output of AGC. As the ACE is driven to zero by the AGC, both frequency and tie-line power errors will be forced to zeros[2]. AGC function can be viewed as a supervisory control function which attempts to match the generation trend within an area to the trend of the randomly changing load of the area, so as to keep the system frequency and the tie-line power flow close to scheduled value. The growth in size and complexity of electric power systems along with an increase in power demand has necessitated the use of intelligent systems that combine knowledge, techniques and methodologies from various sources for the real-time control of power systems.

Researchers all over the world are trying to understand several strategies for AGC of power systems in order to maintain the system frequency and tie line flow at their scheduled values during normal operation and also during small perturbations. A critical literature review on the AGC of power systems has been presented in [3] where various control aspects concerning AGC problem have been studied. Moreover the authors have reported various AGC schemes, AGC strategies and AGC system incorporating BES/SMES, wind turbines, FACTS devices and PV systems. There has been a considerable research work attempting to propose better

AGC systems based on modern control theory [4] and [5], neural network [6], [7], [8] and [9], fuzzy system theory[10], [11] and [12], reinforcement learning [13] and ANFIS approach [14] and [15]. From the literature survey, it may be concluded that there is still scope of work on the optimization of controller parameters to further improve the AGC performance. For this, various novel evolutionary optimization techniques can be proposed and tested for comparative optimization performance study. New artificial intelligence-based approaches have been proposed recently to design a controller. These approaches include particle swarm optimization [16] and [17], differential evolution [18] and [19], multi-objective evolutionary algorithm [20], NSGA-II [21] and [22], etc. Nanda et al. [23]have demonstrated that bacterial foraging, a powerful evolutionary computational technique, based integral controller provides better performance as compared to that with integral controller based on classical and GA techniques in a three unequal area thermal system. Ali and Abd-Elazim [24] have reported recently that Bacterial Foraging Optimization Algorithm (BFOA), based Proportional Integral (PI) controller provides better performance as compared to that with GA based PI controller in two area non-reheat thermal system. Gozde and Taplamacioglu [25] proposed a gain scheduling PI controller for an AGC system consisting of two area thermal power system with governor dead-band nonlinearity. The authors have employed a Craziness based Particle Swarm Optimization (CRAZYPSO) with different objective functions to minimize the settling times and standard error criteria.

Differential Evolution (DE) is a branch of evolutionary algorithms developed by Stron and Price in 1995 for optimization problems [26]. It is a population-based direct search algorithm for global optimization capable of handling non-differentiable, non-linear and multi-modal objective functions, with few, easily chosen, control parameters. It has demonstrated its usefulness and robustness in a variety of applications such as, Neural network learning, Filter design and the optimization of aerodynamics shapes. DE differs from other Evolutionary Algorithms (EA) in the mutation and recombination phases. DE uses weighted differences between solution vectors to change the population whereas in other stochastic techniques such as Genetic Algorithm (GA) and Expert

Systems (ES), perturbation occurs in accordance with a random quantity. DE employs a greedy selection process with inherent elitist features. Also it has a minimum number of EA control parameters, which can be tuned effectively [18] and [19]. In view of the above, an attempt has been made in this paper for the optimal design of DE based PI/PID controller for LFC in two area interconnected power system considering the governor dead-band nonlinearity. The design problem of the proposed controller is formulated as an optimization problem and DE is employed to search for optimal controller parameters. By minimizing the proposed objective functions, in which the deviations in the frequency and tie line power and settling times are involved; dynamic performance of the system is improved. Simulation results are presented to show the effectiveness of the proposed controller in providing good damping characteristic to system oscillations over a wide range of loading conditions, disturbance and system parameters. Further, the superiority of the proposed design approach is illustrated by comparing the proposed approach with a recently published CPSO approach [25] for the same AGC system.

SYSTEM UNDER STUDY

The Automatic Generation Control (AGC) provides the control only during normal changes in load which are small and slow. So the nonlinear equations which describe the dynamic behavior of the system can be linearized around an operating point during these small load changes and a linear model can be used for the analysis thus making the analysis simpler. The system under investigation consists of a two area interconnected power system of thermal plant as shown in Fig. 1. The system is widely used in the literature for the design and analysis of automatic load frequency control of interconnected areas [25]. In Fig. 1, B_1 and B_2 are the frequency bias parameters; ACE_1 and ACE_2 are area control errors; u_1 and u_2 are the control outputs from the controller; R_1 and R_2 are the governor speed regulation parameters in p.u. Hz; T_{G1} and T_{G2} are the speed governor time constants in seconds; ΔP_{G1} and ΔP_{G2} are the changes in governor valve positions (p.u.); T_{T1} and T_{T2} are the turbine time constants in seconds; ΔP_{T1} and ΔP_{T2} are the changes in turbine output powers; ΔP_{D1} and ΔP_{D2} are the load demand changes; ΔP_{Tie} is the incremental change in tie line power (p.u.); K_{PS1} and K_{PS2} are the power system gains; T_{PS1} and T_{PS2} are the power system time constants in seconds; T_{12} is the synchronizing coefficient and Δf_1 and Δf_2 are the system frequency deviations in Hz. The relevant parameters are given in Appendix. The transfer function of governor with non-linearity is given by [25]:

$$G_g = \frac{0.8 - \frac{0.2}{\pi} s}{1 + sT_g} \tag{1}$$

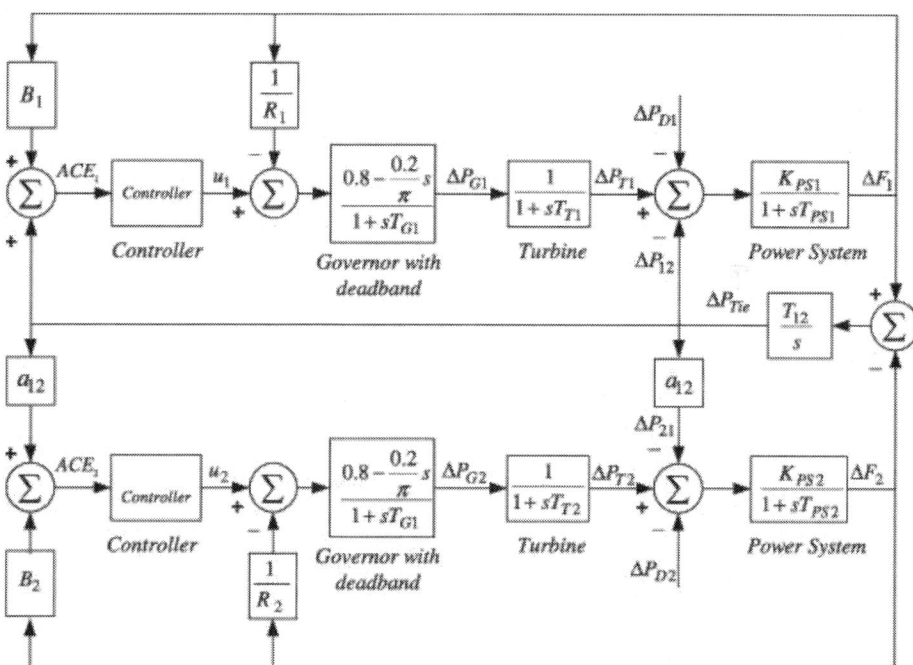

Figure 1. Transfer function model of two-area thermal system with governor dead band.

THE PROPOSED APPROACH

The Proportional Integral Derivative controller (PID) is the most popular feedback controller used in the process industries. It is a robust, easily understood controller that can provide excellent control performance despite the varied dynamic characteristics of the process plant. As the name suggests, the PID algorithm consists of three basic modes, the proportional mode, the integral and the derivative modes. A proportional controller has the effect of reducing the rise time, but never eliminates the steady-state error. An integral control has the effect of eliminating the steady-state error, but it may make the transient response worse. A derivative control has the effect of increasing the stability of the system, reducing the overshoot, and improving the transient response. Proportional Integral (PI) controllers are the most often type used today in industries. A control without Derivative (D) mode is used when: fast response of the system is not required, large disturbances and noises are present during operation of the process and there are large transport delays in the system. Derivative mode improves stability of the system and enables increase in proportional gain and decrease in integral gain which in turn increases speed of the controller response. PID controller is often used when stability and fast response are required. In view of the above, both PI and PID structured controllers are considered in the present paper. Design of PID controller requires determination of the three main parameters, Proportional gain (K_P), Integral time constant (K_I) and Derivative time constant (K_D). For PI controller K_P and K_I are to be determined. The controllers in both the areas are considered to be identical so that $K_{P1} = K_{P2} = K_P$, $K_{I1} = K_{I2} = K_I$ and $K_{D1} = K_{D2} = K_D$.

The error inputs to the controllers are the respective area control errors (ACE) given by:

$$e_1(t) = ACE_1 = B_1 \Delta f_1 + \Delta P Tie \tag{2}$$

$$e_2(t) = ACE_2 = B_2 \Delta f_2 - \Delta P Tie \tag{3}$$

The control inputs of the power system u_1 and u_2 are the outputs of the controllers. With PI structure ($K_{D1} = K_{D2} = 0$) the control inputs are obtained as:

$$u_1 = K_{P1}ACE_1 + K_{I1}\int ACE_1 \tag{4}$$

$$u_2 = K_{P2}ACE_2 + K_{21}\int ACE_2 \tag{5}$$

The control inputs of the power system u_1 and u_2 with PID structure are obtained as:

$$u_1 = K_{P1}ACE_1 + K_{I1}\int ACE_1 + K_{D1}\frac{dACE_1}{dt} \tag{6}$$

$$u_2 = K_{P2}ACE_2 + K_{I2}\int ACE_2 + K_{D2}\frac{dACE_2}{dt} \tag{7}$$

In the design of a PI/PID controller, the objective function is first defined based on the desired specifications and constraints. The design of objective function to tune the controller is generally based on a performance index that considers the entire closed loop response. Typical output specifications in the time domain are peak overshooting, rise time, settling time, and steady-state error. Four kinds of performance criteria usually considered in the control design are the Integral of Time multiplied Absolute Error (ITAE), Integral of Squared Error (ISE), Integral of Time multiplied Squared Error (ITSE) and Integral of Absolute Error (IAE).

Some of the realistic control specifications for Automatic Generation Control (AGC) are[2]:

1. The frequency error should return to zero following a load change.
2. The integral of frequency error should be minimum.
3. The control loop must be characterized by a sufficient degree of stability.
4. Under normal operating conditions, each area should carry its own load.

To meet the above design specifications, three different objective functions are employed in the present paper as given by Eqs. (8), (9) and (10). The first and second objective functions consider only ISE and ITSE criteria given by Eqs. (8) and (9)respectively. The third objective function aims to minimize the ITSE as given by Eq. (9). The third objective function J_3 tries to minimize the settling times of Δf_1, Δf_2 and ΔP_{Tie} in addition to the minimization of all the conventional integral based error criteria.

$$J_1 = ISE = \int_0^{t_{sim}} (\Delta f_1)^2 + (\Delta f_2)^2 + (\Delta P_{Tie})^2 \cdot dt$$

(8)

$$J_2 = ITSE = \int_0^{t_{sim}} [(\Delta f_1)^2 + (\Delta f_2)^2 + (\Delta P_{Tie})^2] \cdot t \cdot dt$$

(9)

$$J_3 = \omega_1 \cdot ISE + \omega_2 \cdot ITSE + \omega_3 \cdot ITAE + \omega_4 \cdot IAE + \omega_5 \cdot T_S \qquad (10)$$

where Δf_1 and Δf_2 are the system frequency deviations; ΔP_{Tie} is the incremental change in tie line power; t_{sim} is the time range of simulation; T_S is the sum of the settling times of frequency and tie line power deviations; ω_1–ω_5 are weighting factors. Inclusion of appropriate weighting factors to the right hand individual terms helps to make each term competitive during the optimization process. Wrong choice of the weighting factors leads to incompatible numerical values of each term involved in the definition of fitness function which gives misleading result. The weights are so chosen that numerical value of all the terms in the right hand side of equation 100 lie in the same range. Repetitive trial run of the optimizing algorithms reveals that numerical value of ISE lies in the range 0.0002–0.02, ITSE value lies in the range 0.002–0.01, ITAE value lies in the range 0.04–1, IAE value lies in the range 0.03–0.5 and total settling times of Δf_1, Δf_2 and ΔP_{Tie} lie in the range 15–50. To make each term competitive during the optimization process the weights are chosen as: $\omega_1 = 10{,}000$, $\omega_2 = 1000$, $\omega_3 = 50$, $\omega_4 = 70$ and $\omega_5 = 0.1$.

The problem constraints are the controller parameter bounds. Therefore, the design problem can be formulated as the following optimization problem:

$$\text{Minimize } J \qquad (11)$$

Subject to

$$K_{P\min} \leqslant K_P \leqslant K_{P\max}, K_{I\min} \leqslant K_I \leqslant K_{I\max} \text{ and } K_{D\min} \leqslant K_D \leqslant K_{D\max} \qquad (12)$$

where J is the objective function $(J_1, J_2, \text{ and } J_3)$ and $K_{P\min}, K_{I\min}; K_{P\max}, K_{I\max}$ and $K_{D\max}, K_{D\max}$ are the minimum and maximum values of the control parameters. As reported in the literature, the minimum and maximum values of controller parameters are chosen as -1.0 and 1.0 respectively.

OPTIMIZATION TECHNIQUE

Differential evolution

Differential Evolution (DE) algorithm is a population-based stochastic optimization algorithm recently introduced [26]. Advantages of DE are: simplicity, efficiency and real coding, easy use, local searching property and speediness. DE works with two populations; old generation and new generation of the same population. The size of the population is adjusted by the parameter N_P. The population consists of real valued vectors with dimension D that equals the number of design parameters/control variables. The population is randomly initialized within the initial parameter bounds. The optimization process is conducted by means of three main operations: mutation, crossover and selection. In each generation, individuals of the current population become target vectors. For each target vector, the mutation operation produces a mutant vector, by adding the weighted difference between two randomly chosen vectors to a third vector. The crossover operation generates a new vector, called trial vector, by mixing the parameters of the mutant vector with those of the target vector. If the trial vector obtains a better fitness value than the target vector, then the trial vector replaces the target vector in the next generation. The evolutionary operators are described below[18] and [19].

Initialization

For each parameter j with lower bound X_j^L and upper bound X_j^U, initial parameter values are usually randomly selected uniformly in the interval $\left[X_j^L, X_j^U\right]$.

Mutation

For a given parameter vector $X_{i,G}$, three vectors $(X_{r1,G}, X_{r2,G},$ and $X_{r3,G})$ are randomly selected such that the indices i, $r1$, $r2$ and $r3$ are distinct. A donor vector $V_{i,G+1}$ is created by adding the weighted difference between the two vectors to the third vector as:

$$V_{i,G+1} = X_{r1,G} + F \cdot (X_{r2,G} - X_{r3,G}) \qquad (13)$$

where F is a constant from $(0, 2)$.

Crossover

Three parents are selected for crossover and the child is a perturbation of one of them. The trial vector $U_{i,G+1}$ is developed from the elements of the target vector $(X_{i,G})$ and the elements of the donor vector $(X_{i,G})$. Elements of the donor vector enter the trial vector with probability CR as:

$$U_{j,i,G+1} = \begin{cases} U_{j,i,G+1} & if \quad rand_{j,i} \leqslant CR \quad or \quad j = I_{rand} \\ X_{j,i,G} & if \quad rand_{j,i} > CR \quad or \quad j \neq I_{rand} \end{cases}$$

$$\qquad (14)$$

With $rand_{j,i} \sim U(0, 1)$, I_{rand} is a random integer from $(1, 2, ..., D)$ where D is the solution's dimension i.e. number of control variables. I_{rand} ensures that $U_{i,G+1} \neq X_{i,G}$.

Selection
The target vector $X_{i,G}$ is compared with the trial vector U_{iG+1} and the one with the better fitness value is admitted to the next generation. The selection operation in DE can be represented by the following equation:

$$X_{i,G+1} = \begin{cases} U_{i,G+1} & \text{if } f(U_{i,G+1}) < f(X_{i,G}) \\ X_{i,G} & \text{otherwise} \end{cases}$$

(15)

where $i \in [1, N_P]$.

Craziness based Particle Swarm Optimization (CPSO)

Particle Swarm Optimization (PSO) is a population based search algorithm for solving the optimization problems. In PSO each individual is referred to as particle and represents a candidate solution. The particles fly through the search space with an adaptable velocity that is dynamically modified according to its own flying experience and also to the flying experience of the other particles. In the original PSO algorithm, the modified velocity and position of each particle are calculated as [16]:

$$v_i^{t+1} = v_i^t + c_1 r_1 \left(p_i^t - x_i^t\right) + c_2 r_2 \left(g^t - x_i^t\right)$$

(16)

$$x_i^{t+1} = x_i^t + v_i^{t+1}$$

(17)

where x_i is the position of ith particle of the swarm, v_i is the velocity of ith particle, n is number of particles in the swarm, t is the number of iterations, c_1 and c_2 are cognitive and social acceleration factors respectively, r_1 and r_2 are random numbers uniformly distributed in the range (0, 1), p_i represents the best previous position of the ith particle, and g represents the best particle among all the particles in the swarm.

The standard PSO algorithm may be trapped in local optima especially for complex problems with many local optima and variables. The Craziness based PSO (CPSO) algorithm can prevent the swarm from being trapped in local minimum, which would cause a premature convergence and lead to fail in finding the global optimum. In the CPSO algorithm, the velocity and position update formula is given by [25]:

$$v_i^{t+1} = r_2 f(r_3)v_i^t + (1 - r_3)c_1 r_1 \left(p_i^t - x_i^t\right) + (1 - r_2)c_2(1 - r_1)\left(g^t - x_i^t\right) \tag{18}$$

$$x_i^{t+1} = x_i^t + v_i^{t+1} + P(r_4)f(r_4)V_{cr} \tag{19}$$

where r_1–r_4 are random numbers uniformly distributed in the range (0, 1), f is a sign function which assigns negative values to r_3 and r_4 if they are less than 0.05 and 0.5 respectively, V_{cr} is a craziness vector linearly decreasing from 10 to 1, $P(r_4)$ is taken as r_4 if r_4 is less than P_{cr}, a predefined probability of craziness, otherwise $P(r_4)$ is taken as zero.

RESULTS AND DISCUSSIONS

Application of DE

The model of the system under study has been developed in MATLAB/SIMULINK environment and DE program has been written (in .mfile). The developed model is simulated in a separate program (by .m file using initial population/controller parameters) considering a 1% step load change in area 1. The objective function is calculated in the .m file and used in the optimization algorithm. The process is repeated for each individual in the population. Using the objective function values, the population is modified by DE for the next generation.

Implementation of DE requires the determination of six fundamental issues: DE step size function also called scaling factor (F), crossover probability (CR), the number of population (N_P), initialization, termination and evaluation function. The scaling factor is a value in the range (0, 2) that controls the amount of perturbation in the mutation process. Crossover probability (CR) constants are generally chosen from the interval (0.5, 1). If the parameter is co-related, then high value of CR work better, the reverse is true for no correlation [18] and [19]. DE offers several variants or strategies for optimization denoted by DE/x/y/z, where x = vector used to generate mutant vectors, y = number of difference vectors used in the mutation process and z = crossover scheme used in the crossover operation. In the present study, a population size of $N_P = 50$, generation number $G = 100$, step size $F = 0.8$ and crossover probability of $CR = 0.8$ have been used. The strategy employed is: DE/best/1/exp. Optimization is terminated by the prespecified number of generations for DE. One more important factor that affects the optimal solution more or less is the range for unknowns. For the very first execution of the program, a wider solution space can be given and after getting the solution one can shorten the solution space nearer to the values obtained in the previous iteration. Here the upper and lower bounds of the gains are chosen as (1, −1). The flow chart of the DE algorithm employed in the present study is given in Fig. 2. Simulations were conducted on an Intel, core 2 Duo CPU of 2.4 GHz and 2 GB MB RAM computer in the MATLAB 7.10.0.499 (R2010a) environment. The optimization was repeated 20 times and the best final solution among the 20 runs is chosen as proposed controller parameters. The best final solutions obtained in the 20 runs are shown in Table 1.

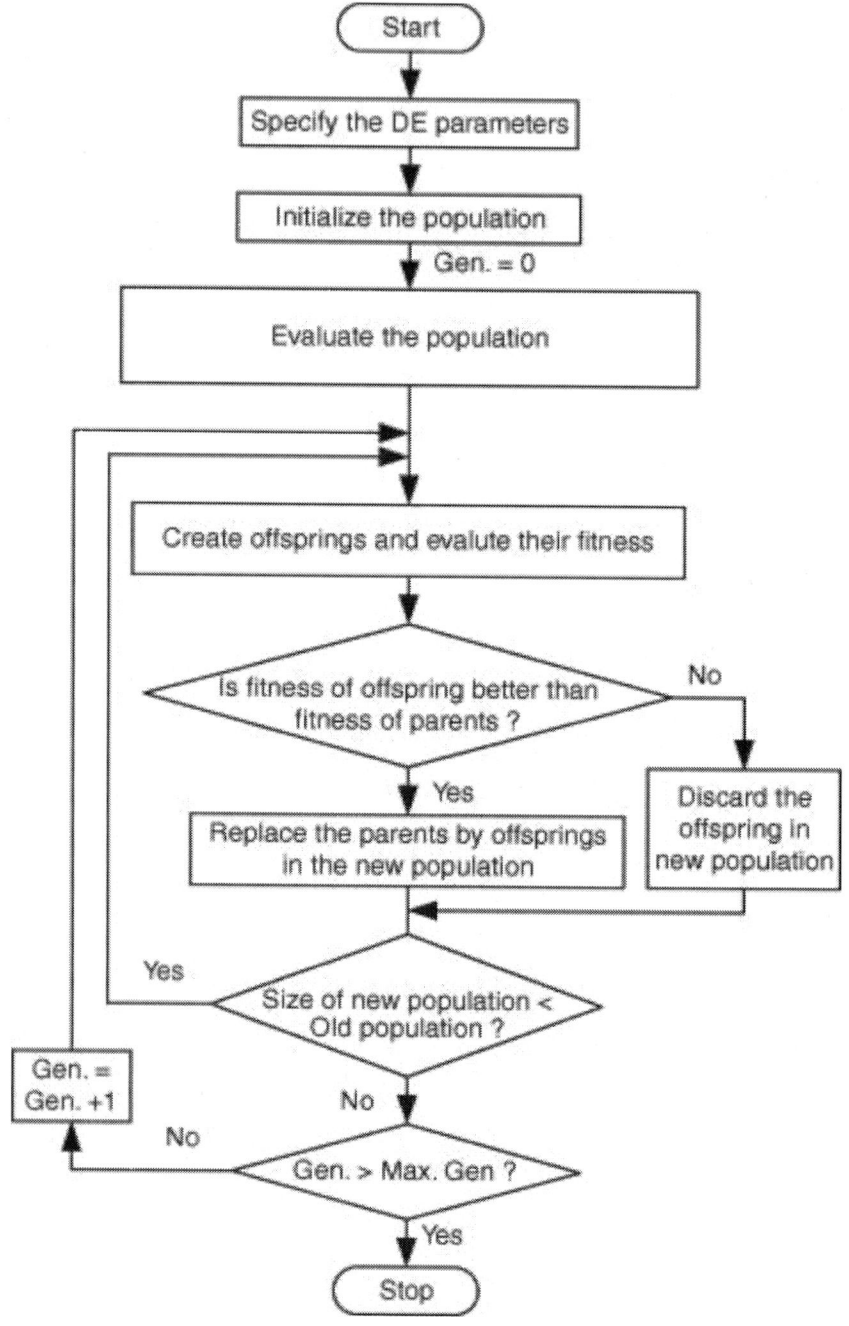

Figure 2. Flow chart of the proposed DE optimization approach.

Table 1. Tuned controller parameters for different objective functions.

Objective function/controller parameters		J_1 (ISE)	J_2 (ITSE)	J_3 (proposed
PI controller	Proportional gain (K_P)	−0.3001	−0.3586	−0.5382
	Integral gain (K_I)	0.3518	0.3159	0.2205
PID controller	Proportional gain (K_P)	0.9041	0.7146	0.2383
	Integral gain (K_I)	0.9322	0.9918	0.9718
	Derivative gain (K_D)	0.9581	0.7595	0.4922

Simulation results

Table 2 shows the ISE value and settling times (2% of final value) when the controller parameters are optimized using ISE error criteria. To show the effectiveness of the proposed DE method results are compared with a recently published CPSO technique for the same interconnected power system and for the same ISE objective function [25]. It can be seen from Table 2 that with the same PI controller structure, the value of ISE obtained using the proposed DE technique is less than that with CPSO technique and minimum ISE is obtained with the DE optimized PID controller. The objective function value is reduced by 5.32% and 88.59% with the proposed DE optimized PI and PID controllers respectively. Also, the settling time for Δf_1 is improved by 0.75% and 69.08% for the proposed PI and PID controllers respectively compared to the results given in [25]. The improvements in settling time for Δf_2 are 4.24% and 72.33% respectively with the proposed PI and PID controllers. For the tie line power deviations ΔP_{tie} the improvements with the proposed PI and PID controllers are 0.71% and 71.75% respectively compared to the CPSO optimized PI controller for the same system.

Table 2. ISE value and settling times with ISE objective function.

Parameters	DE optimized PI controller	DE optimized PID controller	CPSO optimized PI controller[25]
ISE	21.2158×10^{-4}	2.5559×10^{-4}	22.4086×10^{-4}
T_S (s)			
Δf_1	23.66	7.37	23.84
Δf_2	16.82	4.86	17.57
ΔP_{Tie}	23.66	6.73	23.83

The ITSE value and settling times when the controller parameters are optimized using ITSE error criteria are shown in Table 3 along with the CPSO results for the same objective function. It is evident from Table 3 that the better results are obtained with DE compared to CPSO. The improvements are 1.52% and 93.24% in the objective function values with DE optimized PI and PID controllers respectively. For the settling times the improvements are: 28.32% and 70.42% for Δf_1; 25.48% and 64.9%; 28.32% and 70.84% respectively with DE optimized PI and PID controllers.

Table 3. ITSE value and settling times with ITSE objective function.

Parameters	DE optimized PI controller	DE optimized PID controller	CPSO optimized PI controller[25]
ITSE	35.6968×10^{-4}	2.4472×10^{-4}	36.2505×10^{-4}
T_S (s)			
Δf_1	18.83	7.77	26.27
Δf_2	13.42	6.23	18.01
ΔP_{Tie}	18.83	7.66	26.27

To further improving the settling times the proposed objective function J_3 is used and the results are summarized in Table 4. All the four error values and the settling times are compared with the best claimed objective function optimized using CPSO [25]. The respective improvements are also given in Table 4 from which it is clear that the proposed DE optimized PI controller outperforms the CPSO optimized PI controller and best performance is obtained with DE optimized PID controller.

Table 4. Error criteria and settling times with the proposed objective function J_3.

Para-meters	DE optimized PI controller		DE optimized PID controller		CPSO optimized PI controller [25]
	Value	Impr-ove-ment (%)	Value	Imp-rove-ment (%)	
ISE	37.4623×10^{-4}	14.47	4.0257×10^{-4}	90.81	43.8016×10^{-4}
ITSE	66.445×10^{-4}	20.12	3.7025×10^{-4}	95.55	83.1849×10^{-4}
ITAE	48.2145×10^{-2}	17.29	7.29×10^{-2}	87.49	58.2969×10^{-2}
IAE	19.4063×10^{-2}	11.28	4.7644×10^{-2}	78.22	21.875×10^{-2}
T_S (s)					
Δf_1	10.67	4.65	6.87	38.61	11.19
Δf_2	9.64	4.98	4.23	62.33	11.23
ΔP_{Tie}	10.36	20.39	5.91	51.19	12.11

The above analysis shows that the system performance is greatly improved by applying the proposed controllers. Time domain simulations are performed for step load change at different locations. A step increase in demand of 1% is applied at $t = 0$ s in area-1. The system dynamic responses with three objective functions (J_1: ISE; J_2: ITSE and J_3: Proposed) are shown in Figure 3, Figure 4, Figure 5, Figure 6, Figure 7, Figure 8, Figure 9, Figure 10 and Figure 11. In all the figures the response with DE optimized PI and PID controllers is shown with dashed lines (legend 'DE PI') and solid lines (legend 'DE PID') respectively. For comparison the simulation results with CPSO optimized PI controller are also shown in Figure 3, Figure 4, Figure 5, Figure 6, Figure 7, Figure 8, Figure 9, Figure 10 and Figure 11 with dotted lines (legend 'CPSO

PI'). Critical analysis of the dynamic responses clearly reveals that dynamic performance of DE PI controller is better than CPSO PI controller and the best performance is obtained with DE PID controller.

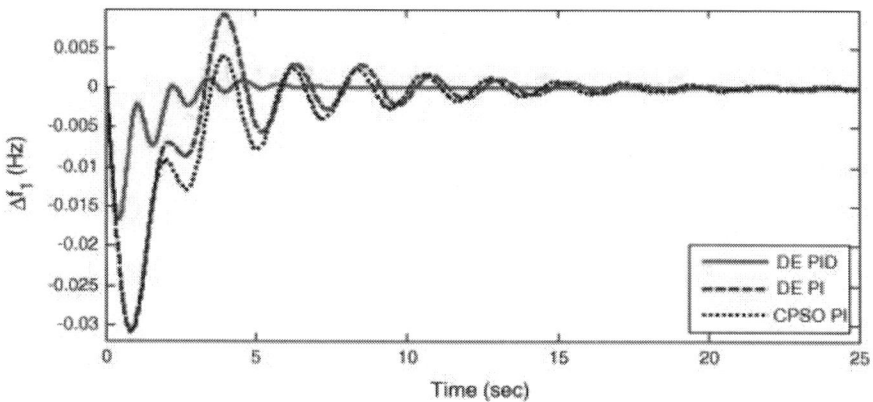

Figure 3. Change in frequency of area-1 for 1% step load increase in area-1 with ISE objective function.

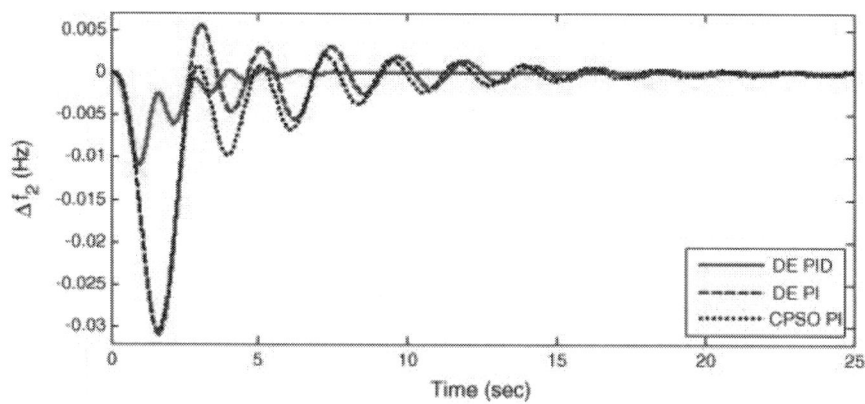

Figure 4. Change in frequency of area-2 for 1% step load increase in area-1 with ISE objective function.

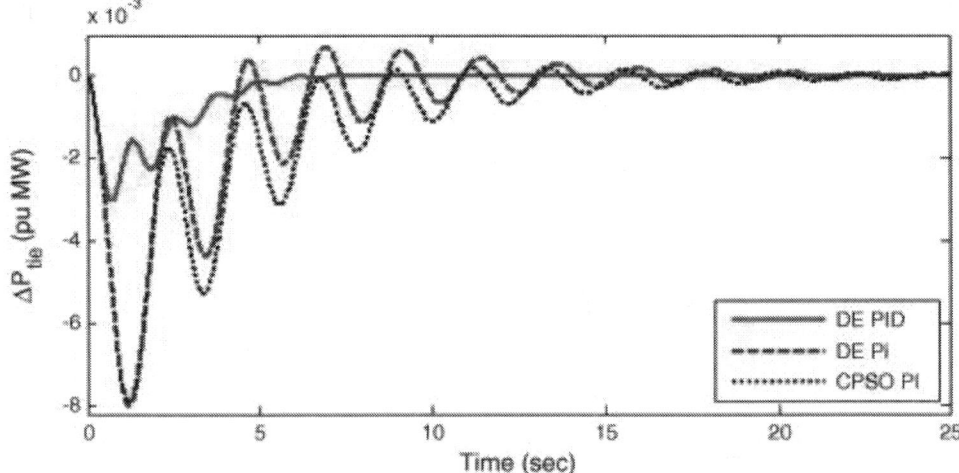

Figure 5. Change in tie line power for 1% step load increase in area-1 with ISE objective function.

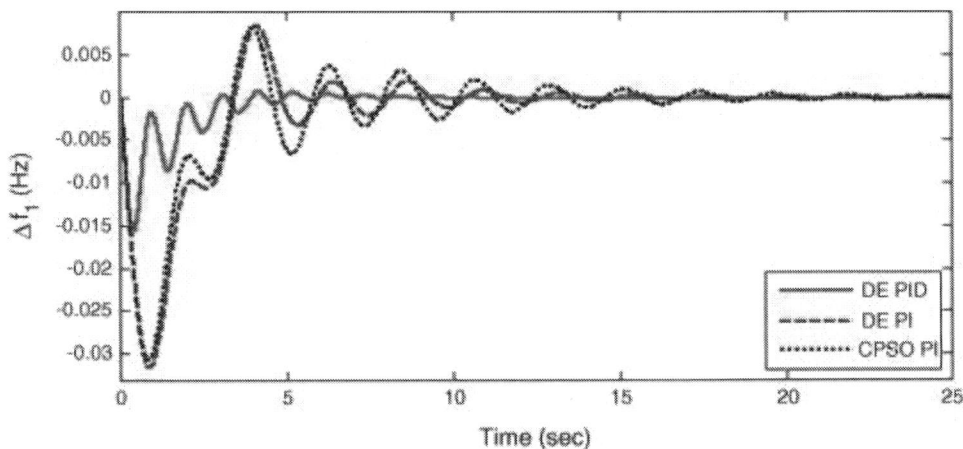

Figure 6. Change in frequency of area-1 for 1% step load increase in area-1 with ITSE objective function.

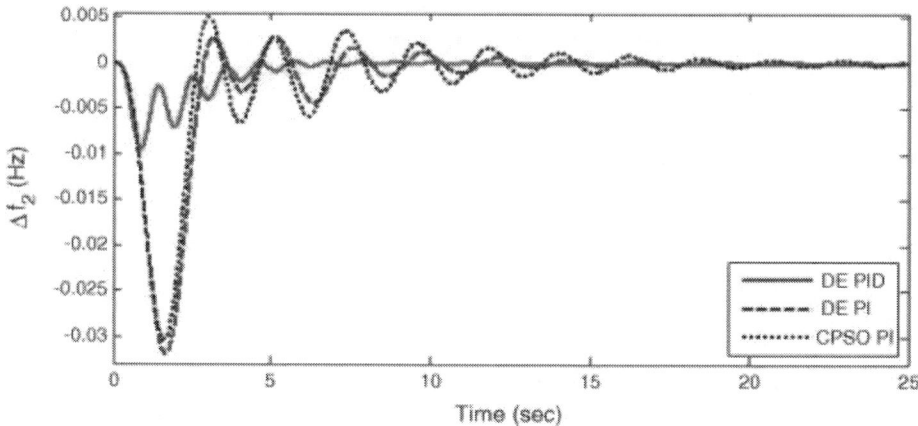

Figure 7. Change in frequency of area-2 for 1% step load increase in area-1 with ITSE objective function.

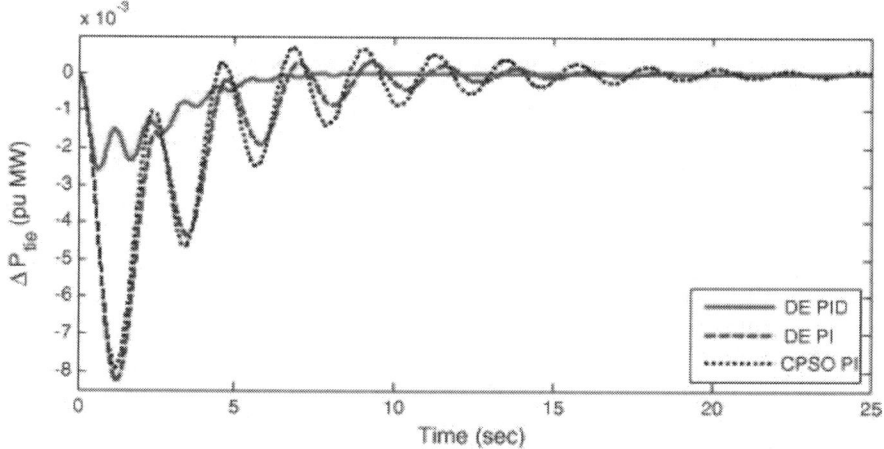

Figure 8. Change in tie line power for 1% step load increase in area-1 with ITSE objective function.

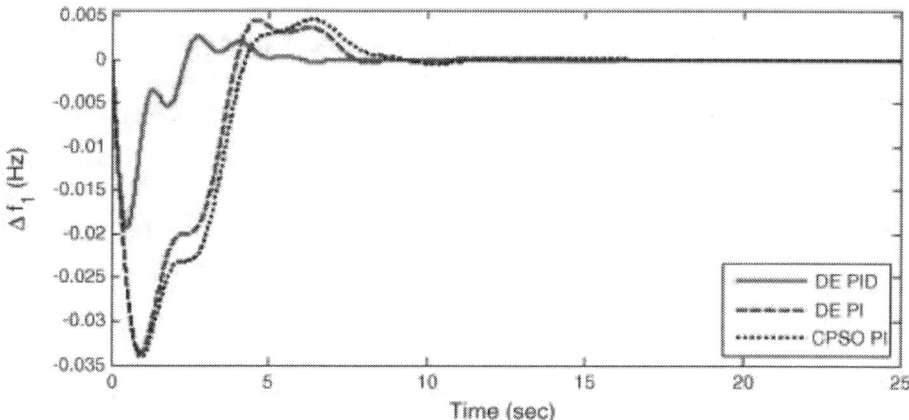

Figure 9. Change in frequency of area-1 for 1% step load increase in area-1 with the proposed objective function.

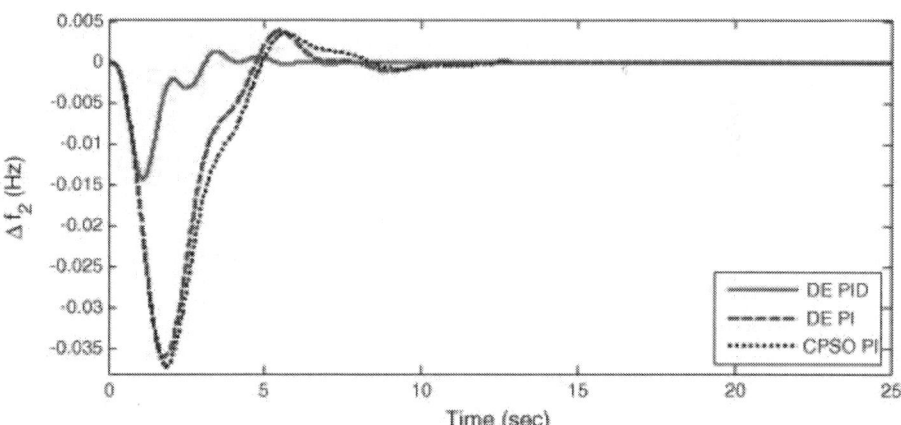

Figure 10. Change in frequency of area-2 for 1% step load increase in area-1 with the proposed objective function.

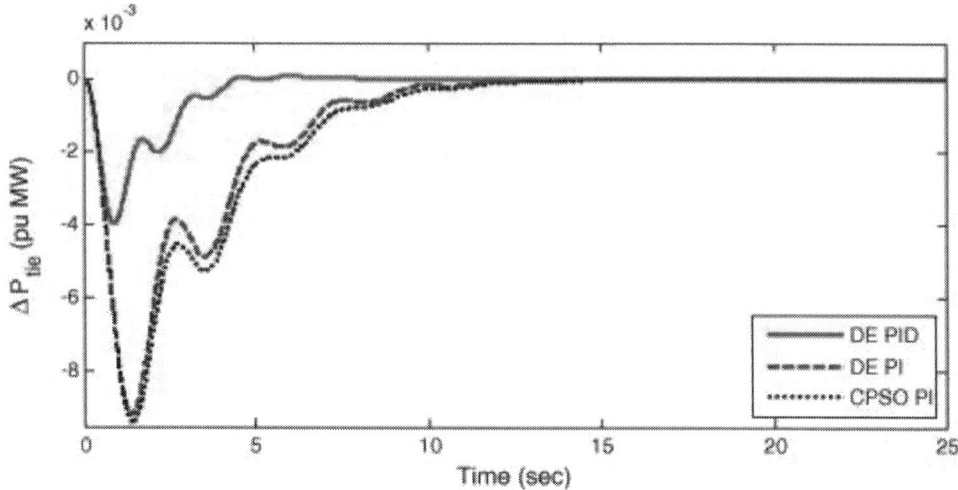

Figure 11. Change in tie line power for 1% step load increase in area-1 with the proposed objective function.

To show the robustness of the control strategy optimized by DE algorithm, controller parameters are tuned at +25%, +50%, −25% and −50% changes in the load demand. As the power exchange between control areas is minimized with the decrease in settling times, the proposed objective function J_3 is used due to its better settling time. The tuned parameters are shown in Table 5. The settling times and its percentage improvements compared to the CPSO technique [25] are given in Table 6. It is clear from Table 6 that settling time is less with DE PI compared to CPSO PI at all the loading conditions and minimum settling times are obtained with DE PID. Figure 12, Figure 13, Figure 14, Figure 15, Figure 16 and Figure 17 show the dynamic response of the system under the above load demand variations. It is clear from Figure 12, Figure 13, Figure 14, Figure 15, Figure 16 and Figure 17 that the designed controllers are robust and perform satisfactorily when load demand changes.

Table 5. Tuned controller parameters for different loadings.

Objective function/controller parameters		25%	50%	−25%	−50%
PI controller	Proportional gain (K_P)	−0.502	−0.5552	−0.5189	−0.5235
	Integral gain (K_I)	0.248	0.2151	0.226	0.2516
PID controller	Proportional gain (K_P)	0.123	0.5602	0.182	0.3599
	Integral gain (K_I)	0.796	0.9628	0.9	0.9349
	Derivative gain (K_D)	0.47	0.6882	0.51	0.6364

Table 6. Settling times at different loadings.

Parameters		DE optimized PI controller		DE optimized PID controller		CPSO optimized PI controller [25]
		Value	Improvement (%)	Value	Improvement (%)	
T_S (s) at +25%	Δf_1	12.14	18.68	6.24	58.2	14.93
	Δf_2	9.34	29.82	4.6	65.44	13.31
	ΔP Tie	10.66	28.55	5.64	62.19	14.92
T_S (s) at +50%	Δf_1	11.05	7.68	6.63	44.61	11.97
	Δf_2	11.36	2.9	5.09	56.49	11.7
	ΔP Tie	11.86	0.16	7.06	40.57	11.88
T_S (s) at −25%	Δf_1	10.23	11.04	5.89	48.78	11.5
	Δf_2	9.43	28.18	4.19	68.08	13.13

T_S (s) at -50%						
	ΔP_{Tie}	9.97	15	5.38	54.13	11.73
	Δf_1	8.97	13.15	5.61	45.74	10.34
	Δf_2	8.58	6.02	3.96	56.62	9.13
	ΔP_{Tie}	9.87	2.75	4.73	53.34	10.15

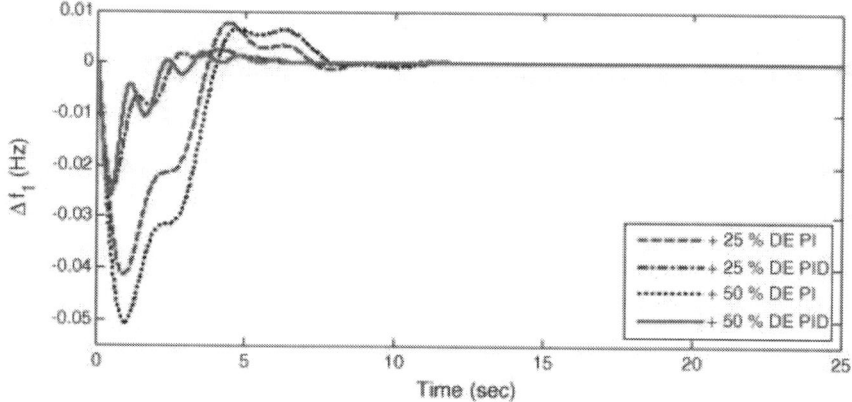

Figure 12. Change in frequency of area-1 for increase in load demands (+25% to +50%) in area-1.

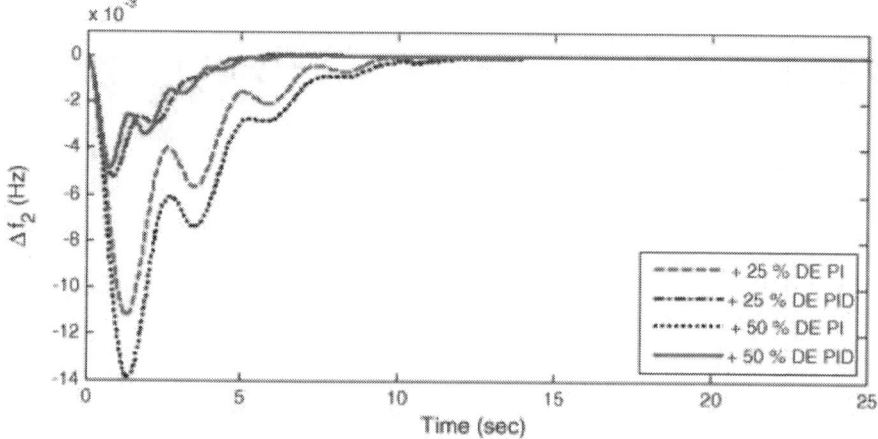

Figure 13. Change in frequency of area-2 for increase in load demands (+25% to +50%) in area-1.

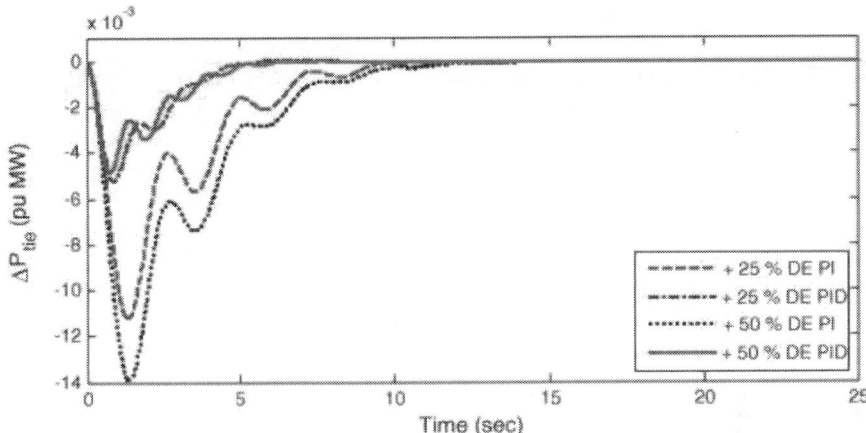

Figure 14. Change in tie line power for increase in load demands (+25% to +50%) in area-1.

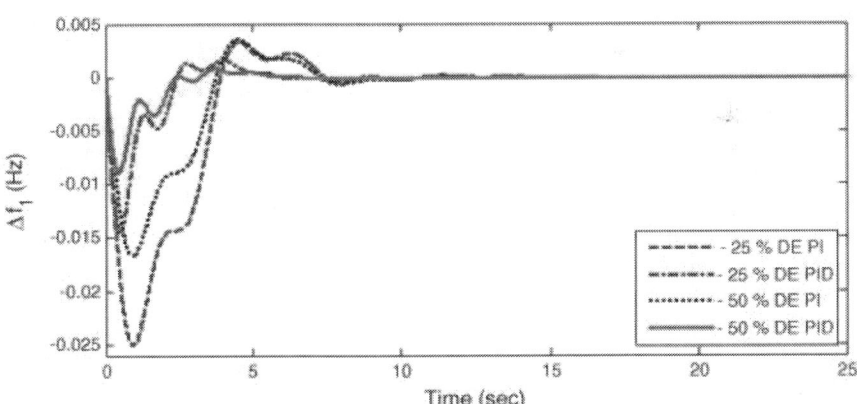

Figure 15. Change in frequency of area-1 for decrease in load demands (−25% to −50%) in area-1.

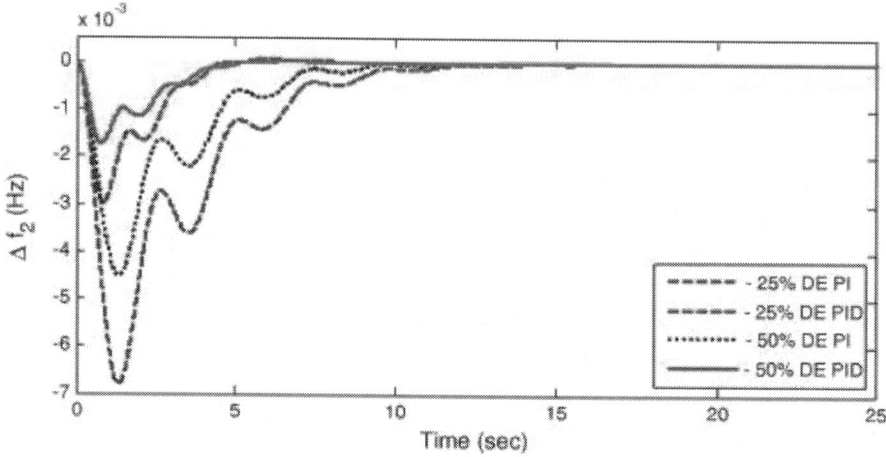

Figure 16. Change in frequency of area-2 for decrease in load demands (−25% to −50%) in area-1.

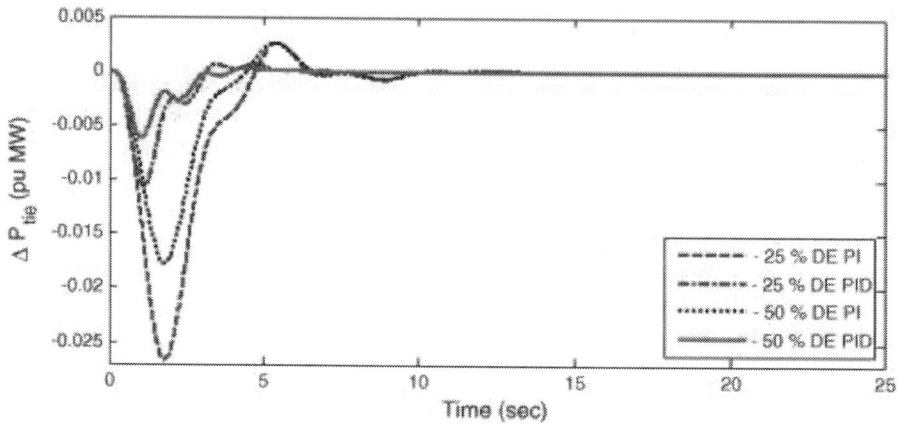

Figure 17. Change in tie line power for decrease in load demands (−25% to −50%) in area-1.

Extension to multi-source system

To get an accurate insight into the AGC topic, it is essential to include the important inherent requirement and the basic physical constraints in the system model. The important constraints which affect the power system performance are boiler dynamics for thermal plants, Generation Rate Constraint (GRC), and Governor Dead Band (GDB) nonlinearity [27]. In view of the above, the study is further extended to a more realistic network of two-area six unit system with different power generating units considering the above physical constraints as shown in Fig. 18. In the first area thermal, hydro and wind generating units are considered and in the second area thermal, hydro and diesel generating units are assumed. The transfer function model of wind and diesel generating units is adopted from [28]. The transfer function model of wind turbine system with pitch control is shown in Fig. 18. The model consists of a hydraulic pitch actuator, data fit pitch response and blade characteristics. The diesel unit is represented by a transfer function as shown in Fig. 18. Each unit has its regulation parameter and participation factor which decide the contribution to the nominal loading, summation of participation factor of each control being equal to 1. Participation factors for thermal and hydro are assumed as 0.575 and 0.3 respectively. For wind and diesel same participation factors of 0.125 are assumed. The nominal parameters of the system under study are given in Appendix B.

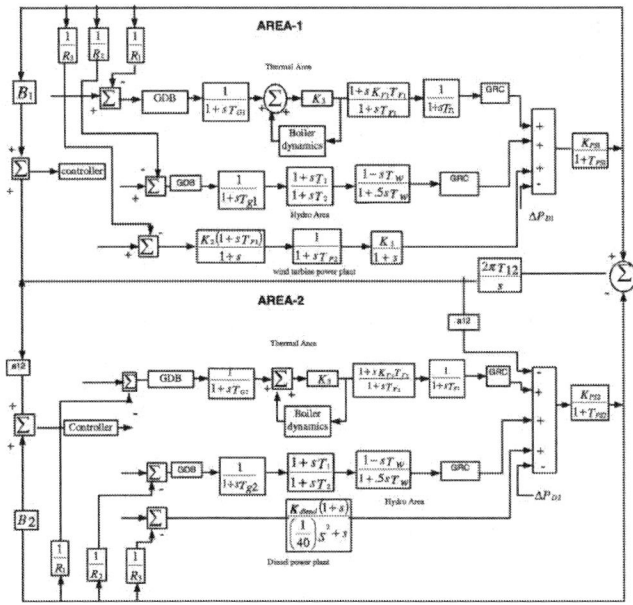

Figure 18. Multi-area multi source power system with nonlinearities.

To include the effect of the boiler dynamics for thermal units, the detailed configuration shown in Fig. 19[29] is considered. This model considers the long-term dynamics of fuel and steam flow on boiler drum pressure as well as combustion controls. Governor dead band is defined as the total amount of a continued speed change within which there is no change in valve position. Steam turbine dead band is due to the backlash in the linkage connecting the servo piston to the camshaft. Much of this appears to occur in the rack and pinion used to rotate the camshaft that operates the control valves. Due to the governor dead band, an increase/decrease in speed can occur before the position of the valve changes. The speed governor dead band has a great effect on the dynamic performance of electric energy system. The backlash non-linearity tends to produce a continuous sinusoidal oscillation with a natural period of about 2 s. For this analysis, in this study backlash non-linearity of 0.05% for the thermal system and 0.02% for the hydro system is considered. In a power system, power generation can change only at a specified maximum rate known as Generation Rate Constraint (GRC). In the present study, a GRC of 3% per min is considered for thermal units. The GRC's for hydro unit are 270% per minute for raising generation and 360% per minute for lowering generation are considered [15]. As the areas are assumed unequal, different PID controllers are considered for each generating unit. To investigate the effect of wind and diesel generation on the system performance, two cases are considered i.e. Case-A: System with thermal and hydro generating units (without wind and diesel units) and Case-B: System with thermal, hydro, wind and diesel generating units. When wind and diesel units are not considered in the system model, the participation factors of thermal unit are increased to 0.695(0.57 + 0.125). The same procedure as described in Section 5.1 is followed to optimize the PID controller parameters of each generating unit in each case. In all the cases, the proposed objective function J_3 given by Eq. (10) is used due to its better performance. The final controller parameters are given in Table 7. A step increase in demand of 1% is applied at $t = 0$ s in area-1 and the system responses are shown inFigure 20, Figure 21 and Figure 22. The settling times and various error criteria for the above case are provided in Table 8. It is clear from Figure 20, Figure 21 and Figure 22and Table 8 that, when the physical constraints are included in the system model, the system performance degrades for Case A i.e. the system with thermal and hydro units. It is also clear from Figure 20, Figure 21 and Figure 22 and Table 8 that the system performance improves with the inclusion of wind and diesel units. The improvements in the system response in Case B are due to the absence of physical constraints for wind and diesel units and they can quickly pick up the additional load demand thus stabilizing the system more quickly. For completeness, a 1% step increase in load demand is applied at $t = 0$ s in area-2 and the system responses are shown in Figure 23, Figure 24 and Figure 25. It is clear from Figure 23, Figure 24 and Figure 25 that the designed controllers are robust and perform satisfactorily when the location of disturbance changes. It is clear from Figure 20, Figure 21, Figure 22, Figure 23, Figure 24 and Figure 25 and Table 8 that the proposed approach can be applied to interconnected power systems with different sources of generation and different PID controllers for each generating unit.

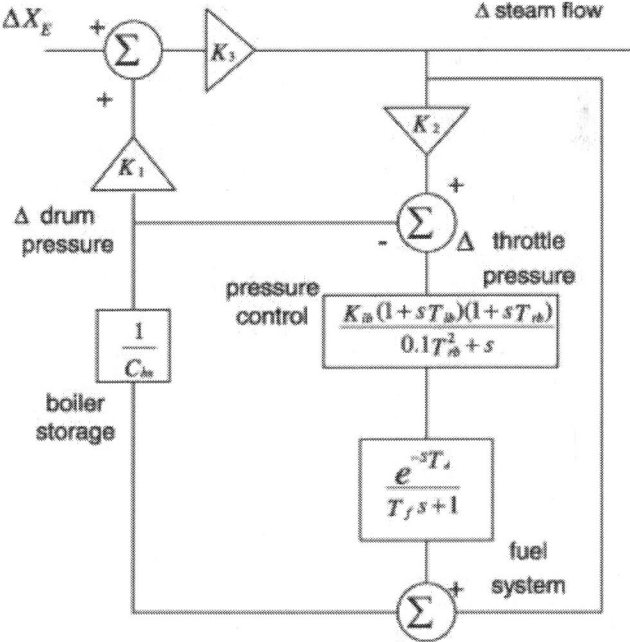

Figure 19. Boiler dynamics configuration.

Table 7. Tuned controller parameters for of multi-source power system with physical constraints.

Parameters/generating units	Case A: System with thermal and hydro units (no wind and diesel)			Case B: System with thermal, hydro, wind and diesel units		
	K_P	K_I	K_D	K_P	K_I	K_D
Area-1						
Thermal	0.705	0.438	0.9681	0.53	0.9272	0.2099
Hydro	0.724	0.657	0.5148	0.982	0.4164	0.2687
Wind	–	–	–	0.731	0.5341	0.3751
Area-2						
Thermal	0.311	0.195	0.151	0.382	0.2172	0.8451
Hydro	0.901	0.071	0.9584	0.072	0.3625	0.1096
Diesel	–	–	–	0.914	0.0405	0.2279

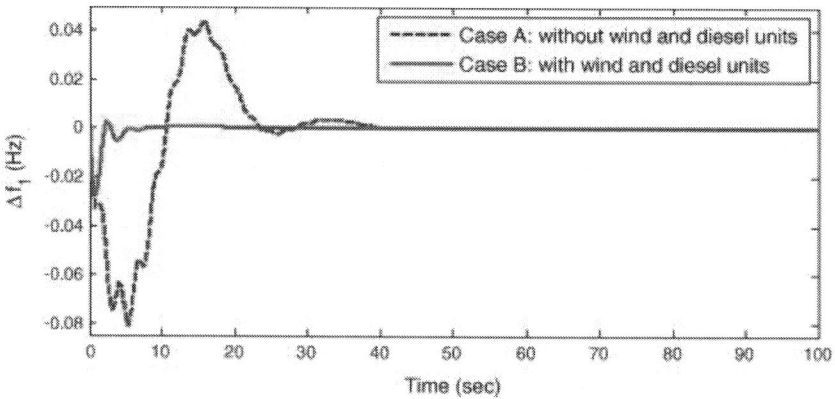

Figure 20. Change in frequency of area-1 for 1% change in area-1 for multisource system with physical constraints.

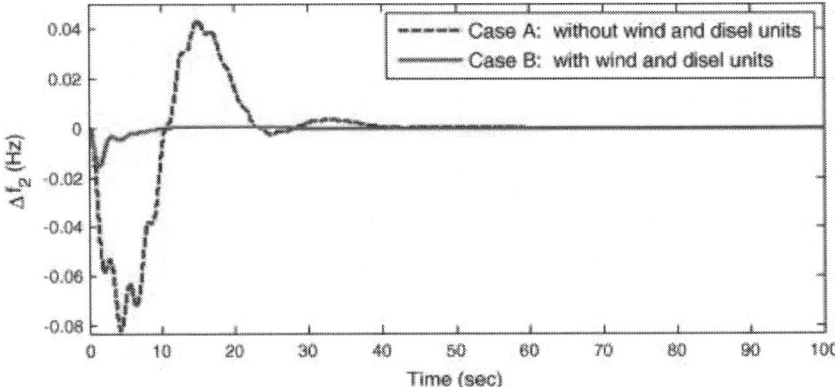

Figure 21. Change in frequency of area-2 for 1% change in area-1 for multisource system with physical constraints.

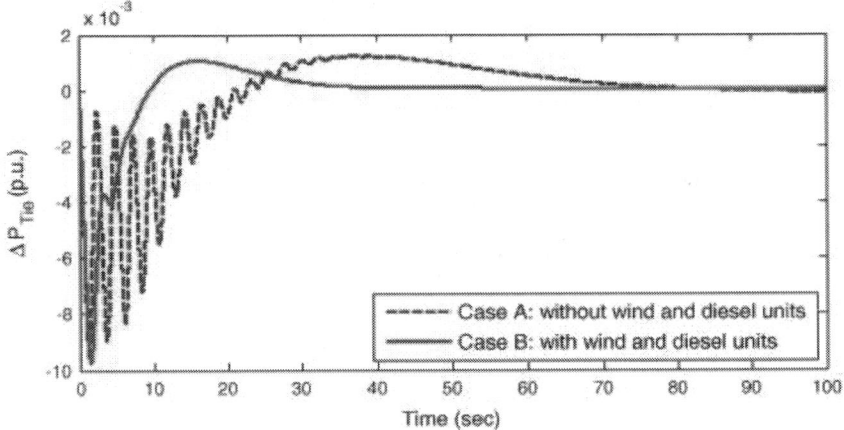

Figure 22. Change in tie-line power for 1% change in area-1 for multisource system with physical constraints.

Table 8. Error criterion and settling times for multi-source power system with physical constraints.

Performance /case	Error criterion				Settling times T_S (s)		
	ISE ($\times 10^{-3}$)	ITSE ($\times 10^{-3}$)	IAE ($\times 10^{-1}$)	ITAE	Δf_1	Δf_2	ΔP_{Tie}
Case A	162.57	1209.7	17.596	20.08	40.6	40.5	60.77
Case B	3.0141	5.554	1.528	1.321	19.7	21.9	25.89

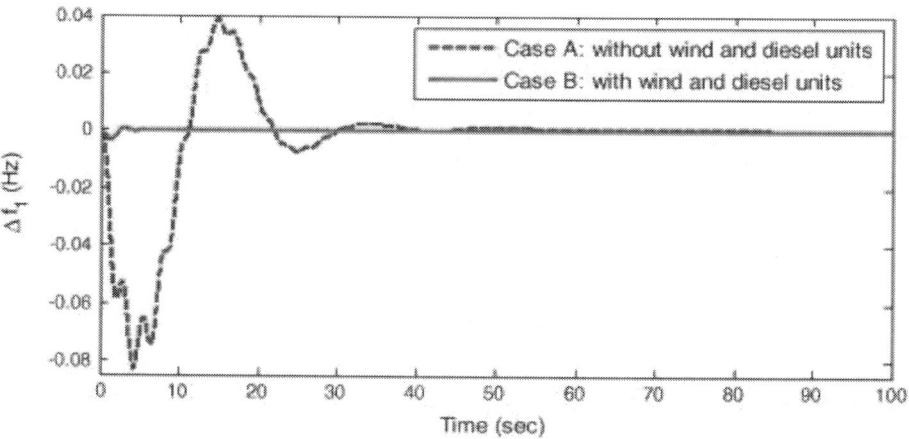

Figure 23. Change in frequency of area-1 for 1% change in area-2 for multisource system with physical constraints.

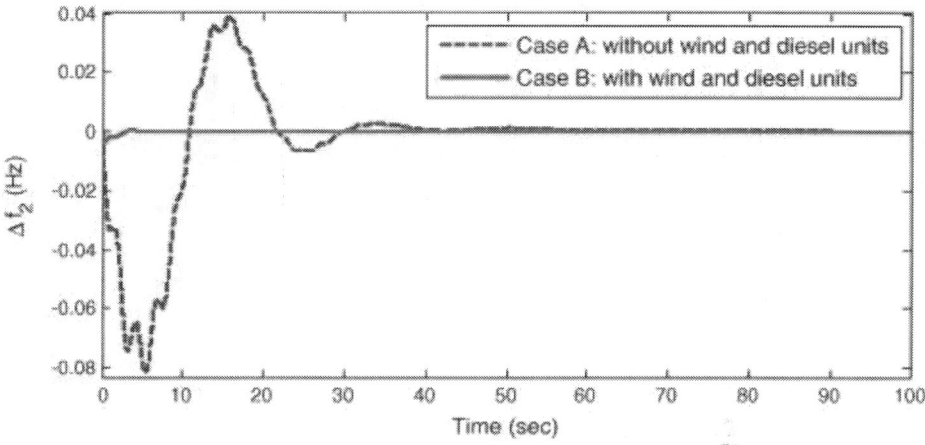

Figure 24. Change in frequency of area-2 for 1% change in area-2 for multisource system with physical constraints.

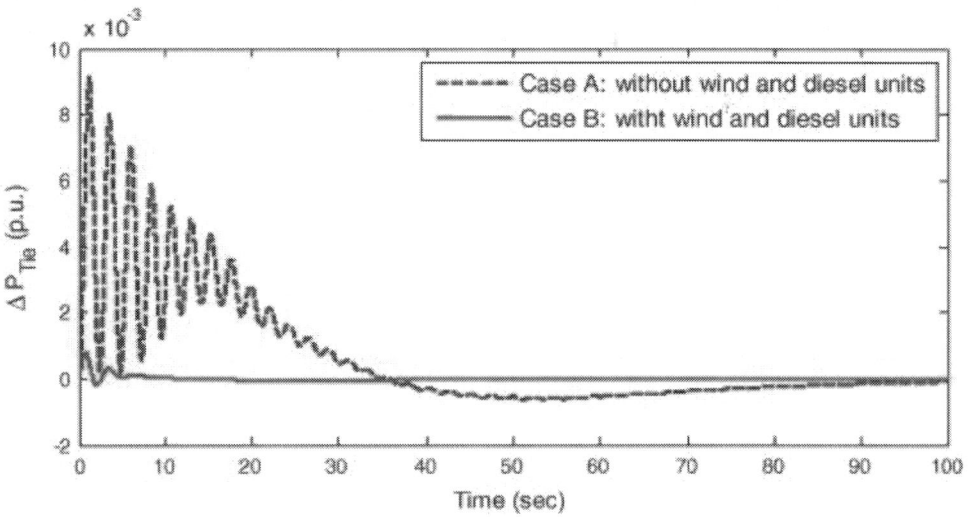

Figure 25. Change in tie-line power for 1% change in area-2 for multisource system with physical constraints.

CONCLUSION

This study presents the design and performance evaluation of Differential Evolution (DE) optimized Proportional–Integral (PI) and Proportional–Integral–Derivative (PID) controllers for Automatic Generation Control (AGC) of an interconnected power system with governor dead-band nonlinearity. For the optimization of controller parameters, selection of suitable objective function is very important. Conventional objective functions used in the literature are Integral of Time multiplied by Squared Error (ITSE), Integral of Squared Error (ISE), Integral of Time multiplied by Absolute Error (ITAE) and Integral of Absolute Error (IAE). Three different objective functions are used for the design purpose in the present paper. The results obtained from the simulations show that the proposed control strategy optimized with a new objective function achieves better dynamic performances than the standard objective functions. The superiority of the proposed approach has been shown by comparing the results with a recently published Craziness based Particle Swarm Optimization (CPSO) technique for the same interconnected power system. It is observed that the proposed DE optimized PI controller outperforms the CPSO optimized PI controller and the best performance is obtained with DE optimized PID controller. Finally, the study is extended to a more realistic network of two-area six unit system with different power generating units considering physical constraints such as boiler dynamics for thermal plants, Generation Rate Constraint (GRC) and Governor Dead Band (GDB) nonlinearity. It is observed that the proposed approach can be applied to interconnected power systems with diverse sources of generation with different PID controllers for each generating unit.

APPENDIX A

B_1, B_2 = 0.425 p.u.
MW/Hz; R_1 = R_2 = 2.4 Hz/p.u.; T_{G1} = T_{G2} = 0.2 s; T_{T1} = T_{T2} = 0.3 s;K_{PS1} = K_{PS2} = 120 Hz/p.u.
MW; T_{PS1} = T_{PS2} = 20 s; T_{12} = 0.0707 p.u.; a_{12} = −1.

APPENDIX B

$B_1 = B_2 = 0.425$ p.u.

MW/Hz; $R_1 = R_2 = 2.4$ Hz/p.u.; $T_{G1} = 0.2$ s; $T_{T1} = 0.3$ s; $T_{G2} = 48.7$ s; T_1 = 0.513 s; $T_2 = 10$ s; $T_w = 1$ s; $T_r = 10$ s; $K_r = 0.333$; $K_1 = 0.85$, $K_2 = 0.09$ 5, $K_3 = 0.92$, $c_b = 200$, $T_d = 0$, $T_f = 10$, $k_{ib} = 0.03$, $T_{ib} = 26$, $T_{rb} = 69$; $K_2 = 1.$ 25; $T_{P2} = 0.041$ s; $K_3 = 1.4$; $T_{P1} = 0.6$ s; $T_1 = 0.025$ s; $K_{PC} = 0.8$; $K_{diesel} = 16.$ 5; $T_{12} = 0.0866$ p.u.

REFERENCES

1. P. Kundur, Power System Stability and control, TMH, 8th reprint, 2009.
2. O.I. Elgerd, Electric Energy Systems Theory. An Introduction, Tata McGraw-Hill, New Delhi, 1983.
3. Ibraheem, P. Kumar, D.P. Kothari, Recent philosophies of automatic generation control strategies in power systems, IEEE Trans. Power Syst. 20 (1) (2005) 346–357.
4. M.L. Kothari, J. Nanda, D.P. Kothari, D. Das, Discrete-mode automatic generation control of a two-area reheat thermal system with new area control error, IEEE Trans. Power Syst. 4 (2) (1989) 730–738.
5. R.R. Shoults, J.A. Jativa Ibarra, Multi area adaptive LFC developed for a comprehensive AGC simulation, IEEE Trans. Power Syst. 8 (2) (1993) 541–547.
6. H.L. Zeynelgil, A. Demiroren, N.S. Sengor, The application of ANN technique to automatic generation control for multi-area power system, Electr. Power Energy Syst. 24 (5) (2002) 345–354.
7. A. Demiroren, H.L. Zeynelgil, N.S. Sengor, Application of ANN technique to load frequency control for three area power system, in: IEEE Power Technol. Conf., Porto, vol. 2, 2001.
8. Q.H. Wu, B.W. Hogg, G.W. Irwin, A neural network regulator for turbo generator, IEEE Trans. Neural Netw. 3 (1) (1992) 95– 100.
9. D.K. Chaturvedi, P.S. Satsangi, P.K. Kalra, Load frequency control: a generalized neural network approach, Electr. Power Energy Syst. 21 (6) (1999) 405–415.

10. S.P. Ghosal, Optimization of PID gains by particle swarm optimization in fuzzy based automatic generation control, Electr. Power Syst. Res. 72 (3) (2004) 203–212.

11. S.P. Ghosal, Application of GA/GA-SA based fuzzy automatic generation control of a multi-area thermal generating system, Electr. Power Syst. Res. 70 (2) (2004) 115–127.

12. J. Talaq, F. Al-Basri, Adaptive fuzzy gain scheduling for load frequency control, IEEE Trans. Power Syst. 14 (1) (1999) 145– 150.

13. T.P. Imthias Ahamed, P.S. Nagendra Rao, P.S. Sastry, A reinforcement learning approach to automatic generation control, Electr. Power Syst. Res. 63 (1) (2002) 9–26.

14. S.H. Hosseini, A.H. Etemadi, Adaptive neuro-fuzzy inference system based automatic generation control, Electr. Power Syst. Res. 78 (7) (2008) 1230–1239.

15. S.R. Khuntia, S. Panda, Simulation study for automatic generation control of a multi-area power system by ANFIS approach, Appl. Soft Comput. 12 (1) (2012) 333–341.

16. S. Panda, N.P. Padhy, Comparison of particle swarm optimization and genetic algorithm for FACTS-based controller design, Appl. Soft Comput. 8 (4) (2008) 1418–1427.

17. S. Panda, N.P. Padhy, Optimal location and controller design of STATCOM for power system stability improvement using PSO, J. Franklin Inst. 345 (2) (2008) 166–181.

18. S. Panda, Differential evolution algorithm for SSSC-based damping controller design considering time delay, J. Franklin Inst. 348 (8) (2011) 1903–1926.

19. Sidhartha Panda, Robust coordinated design of multiple and multi-type damping controller using differential evolution algorithm, Electr. Power Energy Syst. 33 (2011) 1018–1030.

20. S. Panda, Multi-objective evolutionary algorithm for SSSCbased controller design, Electr. Power Syst. Res. 79 (2009) 937– 944.

21. S. Panda, Application of non-dominated sorting genetic algorithm-II technique for optimal FACTS-based controller design, J. Franklin Inst. 347 (7) (2010) 1047–1064.

22. S. Panda, Multi-objective PID controller tuning for a FACTSbased damping stabilizer using non-dominated sorting genetic algorithm-II, Int. J. Electr. Power Energy Syst. 33 (2011) 1296– 1308.

23. J. Nanda, S. Mishra, L.C. Saikia, Maiden application of bacterial foraging based optimization technique in multiarea automatic generation control, IEEE Trans. Power Syst. 24 (2) (2009) 602–609.
24. E.S. Ali, S.M. Abd-Elazim, Bacteria foraging optimization algorithm based load frequency controller for interconnected power system, Electr. Power Energy Syst. 33 (2011) 633–638.
25. H. Gozde, M.C. Taplamacioglu, Automatic generation control application with craziness based particle swarm optimization in a thermal power system, Electr. Power Energy Syst. 33 (2011) 8– 16.
26. R. Stron, K. Price, Differential evolution – a simple and efficient adaptive scheme for global optimization over continuous spaces, J. Global Optim. 11 (1995) 341–359.
27. H. Bevrani, Robust Power System Frequency Control, Springer, 2009.
28. D. Dasa, S.K. Aditya, D.P. Kothari, Dynamics of diesel and wind turbine generators on an isolated power system, Electr. Power Energy Syst. 21 (1999) 183–189.
29. A. Demiroren, E. Yesil, Automatic generation control with fuzzy logic controllers in the power system including SMES units, Electr. Power Energy Syst. 26 (2004) 291–305.

CITATION

Banaja Mohanty, Sidhartha Panda, P.K. Hota, Differential evolution algorithm based automatic generation control for interconnected power systems with non-linearity, Alexandria Engineering Journal, Volume 53, Issue 3, September 2014, Pages 537-552, ISSN 1110-0168, http://dx.doi.org/10.1016/j.aej.2014.06.006.

Chapter 3

A New Approach for Automatic Control Modeling, Analysis and Design in Fully Fuzzy Environment

Walaa Ibrahim Gabr

Department of Electrical Engineering, Benha Engineering Faculty, Benha University, Egypt

ABSTRACT

The paper presents a new approach for the modeling, analysis and design of automatic control systems in fully fuzzy environment based on the normalized fuzzy matrices. The approach is also suitable for determining the propagation of fuzziness in automatic control and dynamical systems where all system coefficients are expressed as fuzzy parameters. A new *consolidity chart* is suggested based on the recently newly developed *system consolidity index* for testing the *susceptibility* of the system to withstand changes due to any system or input parameters changes effects. Implementation procedures are elaborated for the consolidity analysis of existing control systems and the design of new ones, including systems comparisons based on such implementation consolidity results. Application of the proposed methodology is demonstrated through illustrative examples, covering fuzzy impulse response of systems, fuzzy Routh–Hurwitz stability criteria, fuzzy controllability and observability. Moreover, the use of the consolidity chart for the appropriate design of control system is elaborated through handling the stabilization of inverted pendulum through pole placement technique. It is also shown that the regions comparison in consolidity chart is based on type of consolidity region shape such as elliptical or circular, slope or angle in degrees of the centerline of the geometric shape, the centroid of the geometric shape, area of the geometric shape, length of principal diagonals of the shape, and the diversity ratio of consolidity points for each region. Finally, it is recommended that the proposed consolidity chart approach be extended as a unified theory for modeling, analysis and design of continuous and digital automatic control systems operating in fully fuzzy environment.

INTRODUCTION

The majority of the applications of fuzzy theory to automatic control systems are basically directed toward the development of fuzzy logic controllers (FLCs) for linear and nonlinear systems with given or unknown systems' models [1], [2], [3], [4] and [5]. A wide class of these controllers constitutes several components namely the rule-base engine, the fuzzification process, the inference mechanism and the defuzzification process. To avoid defuzzification ambiguities which may arise from more than one crisp output value, some weighted-based techniques are commonly used such as the averaging, the center of gravity (centroid), or the root-sum-square methods [6], [7], [8] and [9]. Other FLCs are based on the conventional fuzzy control (Mamdani Type fuzzy control), fuzzy PID control, neuro-fuzzy control, fuzzy sliding-mode control, adaptive fuzzy control, supervisory fuzzy control, and the Takagi and Sugeno (T–S) model-based fuzzy control [10], [11], [12], [13], [14] and [15].

An important aspect that goes along with the development of FLCs is the solution of the general modeling and analysis aspect of automatic control systems operating in fully fuzzy environment. This is the general case where all inputs and system parameters are fuzzy variables. Two typical examples of linear control systems for the *continuous* and*digital* data cases are shown in Fig. 1. There is a definite need to extend the development of all well proven techniques and methodologies of the deterministic case to the full fuzziness situation. The extension should allow the systematic calculations of propagated fuzziness inside the control system of different configurations and representations.

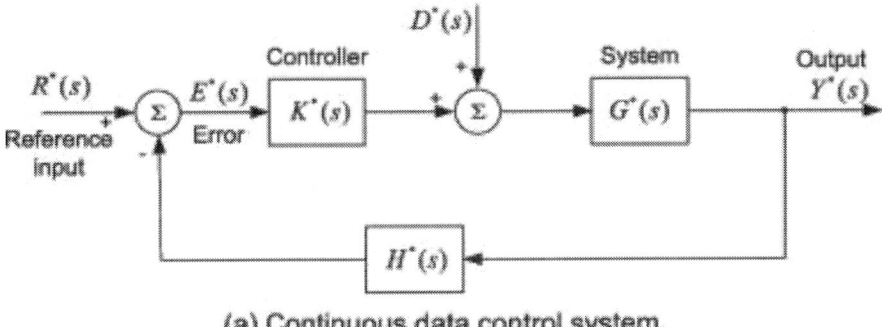

(a) Continuous data control system.

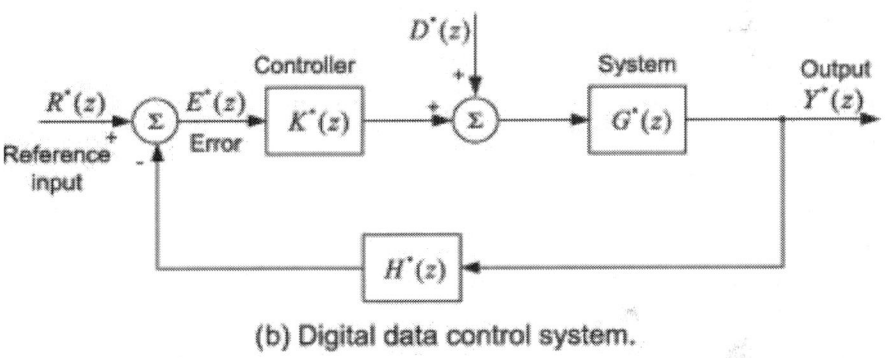

(b) Digital data control system.

( )* means fuzzy parameter.

Figure 1. Two examples of *continuous* and *digital* data control systems operating in fuzzy environments.

The modeling and analysis of automatic control systems operating in fully fuzzy environment is not effectively solved in the literature [16], [17] and [18]. There are many approaches that are carried out to handle this problem using the conventional fuzzy theory. These approaches suffer many drawbacks such as the processing of the solution becomes very cumbersome with the increase of system dimensionality. Moreover, the results obtained by such like approaches are not linear and thus not reversible, leading to that the results obtained in the *forward path* will be different than the *backward path* [19] and [20]. Other techniques, using the direct implementation of fuzzy matrices, also have many shortcomings [21], [22] and [23]. The main hindrance of their spread is heavily related to their impracticability of their present operations (Max, Min, Max.Min, and Min.Max) as they do not reflect any corresponding real-life physical meanings and cause irreversibility and nonlinearity in their processing.

Gabr and Dorrah presented their new notion of Arithmetic and Visual fuzzy logic-based representations [24], [25], [26], [27], [28], [29] and [30]. The approach was based on the normalized fuzzy matrices, where every parameter is expressed by its value and corresponding fuzzy level. It is shown by Gabr [29] and [30] that the proposed Arithmetic fuzzy logic-based representation has corresponding mathematical functions with the conventional fuzzy theory. However, the new approach provides a much easier *arithmetic* rather than *logic* calculations forum that makes its application much practical and effective. Reported methodological experimentations and case studies applications of the Arithmetic and Visual fuzzy logic-based representations were successful in solving some *preliminary* classes of fuzzy global optimization and operations research techniques [24], [25], [26], [27] and [28].

Such fuzziness approach has resulted in the appearance of a new system index named as *"system consolidity index"*. [1]*Consolidity* (the *act* and *quality* of consolidation) is measured by the systems output reactions v ersus combined input/system parameters reaction when subjected to varying environments and events [31], [32], [33], [34],[35] and [36]. Moreover, consolidity can govern the ability of systems to withstand changes when subjected to incurring events or varying environments *"on and above"* normal operation during the system change pathway.

Though the topic of **consolidity theory** has commenced recently in year 2010, there are some good applications of such theory for the analysis and design of automatic control theory. In such applications, the opposite relationship between consolidity and both stability and controllability of state space representation systems was investigated in[33]. Moreover, several examples of applications to automatic control systems were carried out such as the fuzzy design of inverted pendulum using pole replacement method, the optimal design of the fuzzy linear quadratic regulator problem, and the fuzzy Lyapunov stability analysis of the drug concentration control problem [35]. In all these applications, the overall values of consolidity index (average of calculated consolidity points) are only considered in the study without going into any further investigations of the geometric distributions or the diversity analysis of the various consolidity points.

In the following section, the implementation of the consolidity theory is elaborated for further use in the analysis and design of control systems.

METHODOLOGY DEVELOPMENT

Description of the operating fuzzy environments

The fuzzy environments used in system consolidity analysis could be classified as shown in Table 1, and their representations are elucidated in Fig. 2.

Table 1. Various classifications of fuzzy environments for consolidity analysis.

Ser.	Class name	Class abbreviation	Description
1	Open fully fuzzy environment	EO	It is open fully fuzzy environment where all fuzzy levels can equally change all over the positive and negative values of the environment
2	Conditionally open fully fuzzy environment	EC	It is an open fully fuzzy environment but it has the following conditions: (i) The changes of fuzzy levels of parameters may be correlated, and/or (ii) The fuzzy levels may have certain possibility of occurrence similar to Type-2 Fuzzy algebra.

| 3 | Bounded fully fuzzy environment | EB | It is a restricted fully fuzzy environment where all fuzzy levels can equally change all over the positive and negative within restricted ranges in the environment |
| 4 | Partial fuzzy environment | EP | It is a restricted partial fuzzy environment where only some fuzzy (or none fuzzy) levels can differently or conditionally change within restricted ranges in the partial environment |

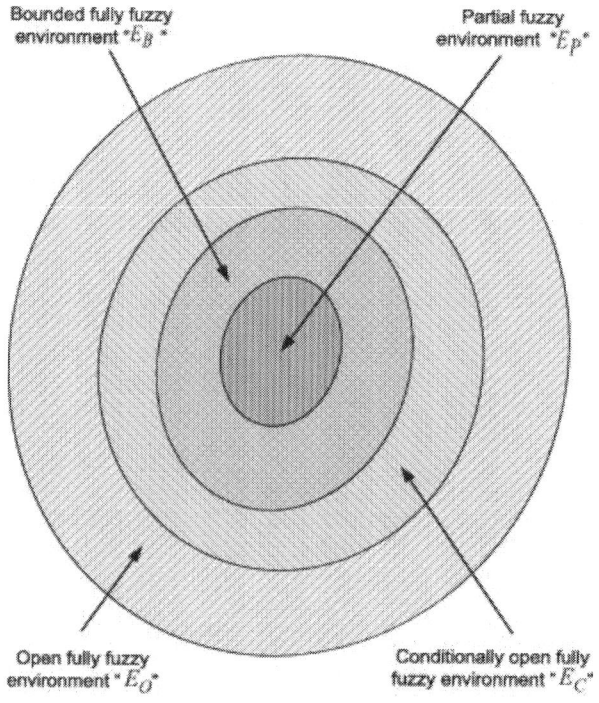

Figure 2. Various classifications of different fuzzy environments.

In order not to complicate the mater unduly in this paper, our analysis will be based on classes $_{EO}$ and $_{EB}$. The system consolidity analysis applies similarly to other fuzzy environment classes.

The consolidity methodology
The consolidity methodology for the analysis and design of control systems is based in modeling the system input and parameters as fuzzy variables, leading to a corresponding output of the similar fuzzy nature. A system operating at a certain stable original state in fully fuzzy environment is said to be *consolidated* if its overall output is suppressed corresponding to their combined input and parameters effect, and vice versa for *unconsolidated* systems. Neutrally consolidated systems correspond to marginal or balanced reaction of output, versus combined input and system.

In general, the output fuzziness behavior toward input fuzziness could differ from one system to another. Examples of these behaviors are as follows:

i.   Outputs could absorb the input fuzziness and give smaller or diminishing output fuzziness.
ii.   Outputs could yield almost the same level of input fuzziness.
iii. Output could give higher output fuzziness compared to input fuzziness.

Let us assume a general system operating in fully fuzzy environment, having the following elements:

Input parameters:

$$\underline{I} = (V_{I_i}, \ell_{I_i})$$

(1)

such that $V_{I_i}, i = 1, 2, \ldots, m$ describe the deterministic value of input component I_i, and fl_i indicates its corresponding fuzzy level.

System parameters:

$$\underline{S} = (V_{S_j}, \ell_{S_j}) \tag{2}$$

such that $V_{S_j}, j = 1, 2, \ldots, n$ denote the deterministic value of system parameter s_j, and fs_j denotes its corresponding fuzzy level.

Output parameters:

$$\underline{O} = (V_{O_i}, \ell_{O_i}) \tag{3}$$

such that $V_{O_j}, i = 1, 2, \ldots, k$ designate the deterministic value of output component o_i, and fo_i designates its corresponding fuzzy level.

We will apply in this investigation, the overall fuzzy levels notion, first for the combined input and system parameters, and second for output parameters. As the relation between combined input and system with output is close to (or of the like type) of the multiplicative relations, the multiplication fuzziness property is applied for combining the fuzziness of input and system parameters.

For the combined input and system parameters, we have the weighted fuzzy level to be denoted as the combined *Input and System Fuzziness Factor*$_{\text{FI+S}}$, given as:

$$F_{1+S} = \frac{\sum_{i=1}^{m} V_{I_i} \cdot \ell_{I_i}}{\sum_{i=1}^{m} V_{I_i}} + \frac{\sum_{j=1}^{n} V_{S_j} \cdot \ell_{S_j}}{\sum_{j=1}^{n} V_{S_j}} \tag{4}$$

Similarly, for the **Output Fuzziness Factor**$_{FO}$, we have

$$F_O = \frac{\sum_{i=1}^{k} V_{O_i} \cdot \ell_{O_i}}{\sum_{i=1}^{k} V_{O_i}}$$

(5)

Let the positive ratio $|FO/FI+S|$ defines the **System Consolidity Index**, to be denoted as$_{FO/(I+S)}$. Based on $_{FO/(I+S)}$ the system consolidity state can then be classified as [20],[21], [22], [23], [24], [25], [26], [27], [28], [29], [30], [31], [32] and [33]:

i.   **Consolidated**if $_{FO/(I+S)}$<1, to be referred to as "*Class C*".
ii. **Neutrally Consolidated**if $_{FO/(I+S)}$≈1, to be denoted by "*Class N*".
iii. **Unconsolidated**if $_{FO/(I+S)}$>1, to be referred to as "*Class U*".

For cases where the system consolidity indices lie at both consolidated and unconsolidated parts, the system consolidity will be designated as a mixed class or "*Class M*".
The selection of the fuzzy levels testing scenarios for both the system and input should follow the same usual consideration. First of all the input and system fuzzy values for system consolidity testing are selected as integer values to be preferably in the range ±8 for open fuzzy environment and in the range ±4 for bounded fuzzy environments. Nevertheless, the output fuzzy level could assume open values beyond these ranges based on the overall consolidity of the system. However, all over implementation procedure in the paper, the exact fraction values of fuzzy levels are preserved during the calculations and are rounded as integer values only at the final results [34].

It is remarked that the typical ranges of the consolidity indices $FO/(I+S)$ based on previous real-life applications are as follows: *very low* (<0.5), *low* (0.5–1.5), *moderate*(1.5–5), *high* (5–15), and *very high* (>15) [34] and [35]. A good practical consolidated system should have the value of consolidity index $FO/(I+S) \leqslant 1.5$.

The consolidity chart

The concept of implementing the consolidity theory to the analysis and design of control system is to plot for each system its consolidity chart defined as the relation between the*Output Fuzziness Factor*$|FO|$ in the vertical axis (y) and the *Input and System Fuzziness Factor*$|FI+S|$ in the horizontal axis (x).

The best way for sketching each system's region in the consolidity chart is to calculate representative points of *output fuzziness factor* (y-axis) versus *input and the system fuzziness factor* (x-axis) and plot all these x–y points first in the chart. The average consolidity index is then calculated based on these points and its value will represent the slope of the center line of the region under study. The boundary of the region can then be sketched around this center line embodying all (or the majority) of these fuzziness x–$y$points.

Examples of the consolidity regions or patterns of various consolidity classes are summarized in Table 2 and sketched in Fig. 3. The shapes of each region are assumed for simplicity of the elliptical. However, other geometric shapes such as the circular one could take place for various applications.

Table 2. Various classifications of system consolidity.

Ser.	Class name	Class abbreviation	Description
1	Consolidated	C	All values of Consolidity Index are less than 1, that is $_{FO/(I+S)}<1$
2	Quasi-consolidated	$\tilde{C}$	A mixed system that is clearly inclined more toward consolidation such as the center of gravity (averaged value) has $_{FO/(I+S)}<1$
3	Neutrally consolidated	N	All values of Consolidity Index are nearly 1, that is $_{FO/(I+S)}\approx1$
4	Mixed class	M	All values of Consolidity Indices lie at both Consolidated and Unconsolidated zones, $_{FO/(I+S)}<$and>1
5	Quasi-unconsolidated	$\tilde{U}$	A mixed system that is clearly inclined more toward un-consolidation such as the center of gravity (averaged value) has $_{FO/(I+S)}>1$
6	Unconsolidated	U	All values of Consolidity Index are more than 1, that is $_{FO/(I+S)}>1$

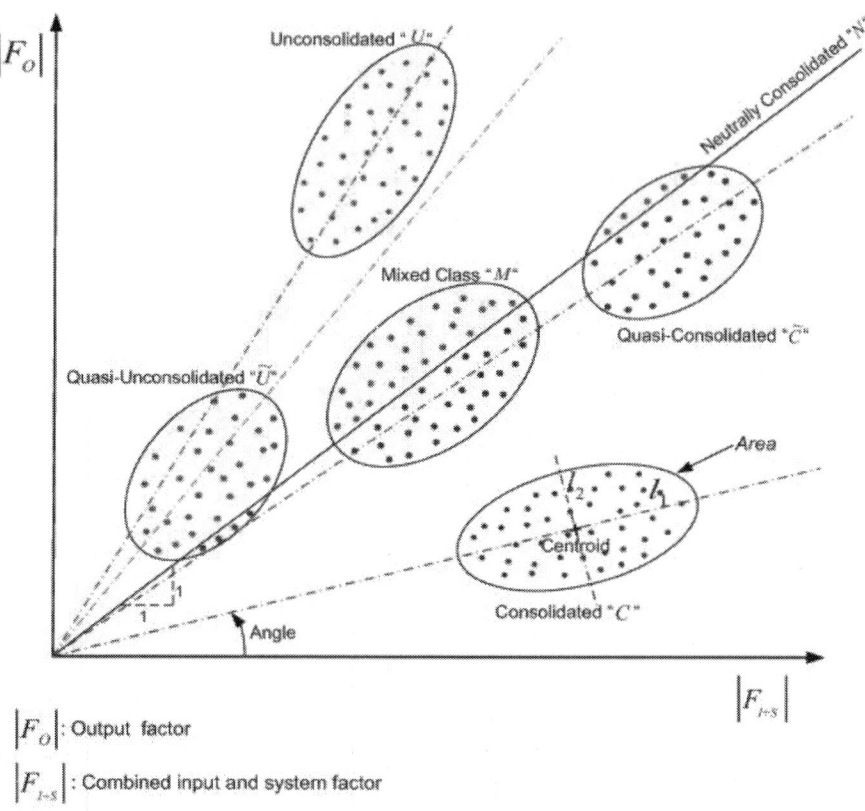

Figure 3. The *consolidity chart* showing six different classes of system consolidity patterns (regions) as described in Table 2.

Based on these consolidity patterns, the degree of susceptibility of the system to withstand the effect of changes in system and input parameters can be evaluated. In fact,*Consolidity index* is an important factor in scaling system parameter changes when subjected to events or varying environment. For instance, for all coming events at say event state μ which are *"on and above"* the system normal situation or stand will lead to consecutive changes of parameters. Such changes follow the general relationship at any event   step μ as: Δ Parameter   change ᵖ = *Function* [consolidity ᵖ, varying environment or event ᵖ] [37] and [38]. Two important common cases in real life of such formulation are the *linear* (or linearized) and the *exponential* relationship.

Implementation approach of consolidity theory to control systems

Implementation approach for existing control systems

For the existing *man-made* or *natural* systems, the situation could be complicated. The testing of these existing systems could reveal the poor consolidity of such system. This is quite expected as we previously used to build all existing systems without considering the new concept of system consolidity. For existing *man-made* systems, the situation could be possible by altering parameters of the system within the utmost extent permitted for changes. As for *natural* systems, the system consolidity improvement matter could also be possible by interfering within the system parameters together with environment and trying to direct the physical process toward better targeted consolidity.

For *existing automatic control systems*, they could be *firstly* examined for their consolidity behavior. Based on the obtained results, appropriate interventions are carried out for adjusting one or more of system key parameters or *possibly* controlling existing operating environment to attain improved consolidity without jeopardizing their stability or performance. This interfering approach for existing systems is illustrated in Fig. 4(a).

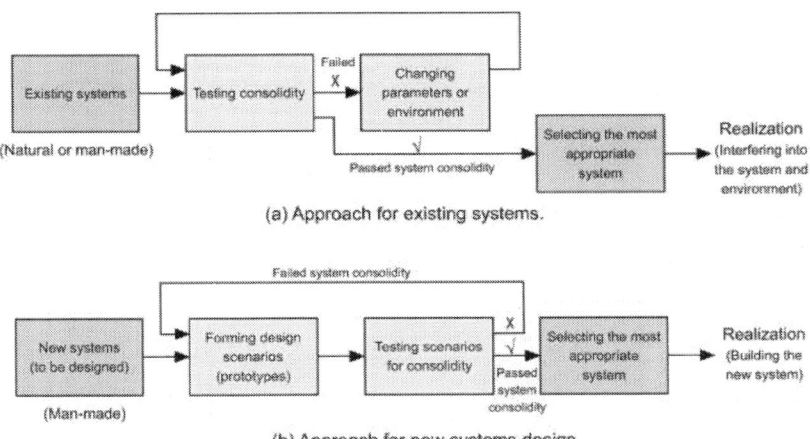

(a) Approach for existing systems.

(b) Approach for new systems design.

Figure 4. Implementation approach of consolidity theory for both existing and new systems under design.

Implementation approach for new control systems design

For *new automatic control systems* design, the implementation of consolidity theory is much simpler. The designers commence using the conventional automatic control techniques leaving at the end one or more flexible (or changeable) parameters that consecutively be adjusted for preserving good system consolidity behavior. Several designs could be developed and then ranked within the framework of consolidity for selecting the best choices that also fulfill acceptable degrees of functionality.

The approach is also applicable for higher dimensional automatic control as it is based mainly on matrix formulations. For design analysis, conventional techniques are first used leaving one or two parameters of flexible ranges. The suggested consolidity technique can then be drawn by varying these parameters to obtain an improved design from the consolidity point of view.

In general, as the generation of these prototypes during the design process is not completely exhaustive. The terms of *superior* or *inferior* of consolidation remain as relative comparison. Such comparison is sufficient for all real-life applications as the system designers could follow later other cycles of improvement to locate a new better superior system that surmounts the old superior design. Such building approach for new systems is elucidated in Fig. 4(b).

It is remarked that during implementation each problem parameters should be addressed first as *symbols* and not be substituted (with fuzziness defined as their pairs or shadows). Conventional mathematics is then applied to the basic variables while the appropriate fuzzy algebra is implemented on their corresponding pairs or shadows (fuzziness). Parameters substitutions are made at the end step of solution leading to the calculation of the consolidity factors as specified by the problem analyst.

For complicated symbolic manipulations (and computations) the use of Matlab Symbolic Toolbox, Mathematica or similar like software libraries could be highly effective to foster the consolidity theory through conducting its necessary derivations. This will enable the implementation of the consolidity analysis to wider classes of linear, nonlinear, multivariable and dynamic problems with different types of complexities.

Systems comparison based on the consolidity indices implementation results

The first step in the design of any specific problem is to carry out the consolidity indices of all various available scenarios. These results are extracted from only one overall consolidity index based on the design philosophy to be followed. Examples of such design basis are given as

(i) *Average consolidity scores*: In this case, the average of all scored consolidity indices for each scenario is calculated and used solely for the selection of the most appropriate consolidated design.

(ii) *Weighted average consolidity scores*: For the situations where the possibility of the input and system fuzziness is known, the weighted average values of the consolidity by these given possibilities are calculated and the results are used in the selection of design.

(iii) *Worst consolidity scores*: In this case, the worst score of the consolidity index is chosen as an overall evaluation index. This could be the maximum (or minimum) of all scored indices if we are seeking the superior (or inferior) consolidated design

Another in-depth direction of comparisons is through plotting regions (patterns) of fuzziness behavior in the sketched consolidity chart similar to the ones shown in Fig. 3. It could appear from such figure that each consolidity region changes from system to another and follows certain geometric shapes such as the elliptical, circulars, or others. The geometric features of each consolidity region can be characterized by the following features:

Symbol	Description
R	Type of consolidity region shape such as elliptical, circular, or other shapes
S	Slope or angle in degrees of the centerline of the geometric shapes of the overall consolidity index S(degrees) = tan^{-1}(overall consolidity index)
$C=(x,y)$	The centroid of the geometric shape expressed by its horizontal coordinate per unit (pu) and the vertical coordinate per unit
A	Area of the geometric shape of R in pu^2
l_1	Length of geometric (major) diagonal in direction of slope of the consolidity region (pu)
l_2	Length of geometric (minor) diagonal in perpendicular to the slope S of the consolidity region (pu)
l_2/l_1	Diversity ratio of consolidity points (unitless)

The comparison or ranking of each consolidity region will be based on less slope, less area and less diversity ratio (l_2/l_1). Moreover, the position of the centroid C=(x,y)(upward or downward) within the geometric shape main centerline depends mainly on the nature of the affected input influences which are particular for each specific application. Higher values of such centers mean higher fuzzy input effects or influences. The above shown features of the consolidity charts will be the basis of the analysis of the various applications given in the following sections of the paper.

METHODOLOGY IMPLEMENTATION TO AUTOMATIC CONTROL SYSTEMS MODELING

Consider the general differential equation [16]

$$\frac{d^n x}{dt^n} + a_{n-1} \cdot \frac{d^{n-1} x}{dt^{n-1}} + \cdots + a_1 \cdot \frac{dx}{dt} + a_0 \cdot x$$

$$= b_{n-1} \cdot \frac{d^{n-1} u}{dt^{n-1}} + \cdots + b_1 \cdot \frac{du}{dt} + b_0 \cdot u \ldots$$

(6)

where all the equation parameters are fuzzy numbers. These fuzzy numbers are expressed by their deterministic values and corresponding fuzzy level as described by the Arithmetic fuzzy logic-based representation.

Define a set of state variables for a typical fuzzy control system as follows:

$$\dot{x}_1 = x_2$$
$$\dot{x}_2 = x_3$$

$$\vdots \quad \vdots$$

$$\dot{x}_n = -a_0 \cdot x_1 - \cdots - a_{n-1} \cdot x_n + u$$

(7)

and an output equation

$$y = b_0' \cdot x_1 + b_1' \cdot x_2 + \cdots + b_{n-1}' \cdot x_n$$

(8)

where $b_0' \ b_1' \ b_2' \ \cdots \ b_{n-1}'$ are fuzzy coefficients.

Then, the state equation is expressed as

$$\begin{bmatrix} \dot{x}_1 \\ \dot{x}_2 \\ \vdots \\ \dot{x}_{n-1} \\ \dot{x}_n \end{bmatrix} = \begin{bmatrix} 0 & 1 & 0 & \cdots & 0 \\ 0 & 0 & 1 & \cdots & 0 \\ \vdots & & & & \\ 0 & 0 & 0 & \cdots & 1 \\ -a_0 & -a_1 & -a_2 & \cdots & -a_{n-1} \end{bmatrix} \begin{bmatrix} x_1 \\ x_2 \\ \vdots \\ x_{n-1} \\ x_n \end{bmatrix} + \begin{bmatrix} 0 \\ 0 \\ \vdots \\ 0 \\ 1 \end{bmatrix} u$$

(9)

The state-space representation of (9) is denoted as the controllable canonical form. The output equation is

$$y = \begin{bmatrix} b'_0 & b'_1 & b'_2 & \cdots & b'_{n-1} \end{bmatrix} \cdot \begin{bmatrix} x_1 \\ x_2 \\ x_3 \\ \vdots \\ x_n \end{bmatrix}$$

(10)

Consider now the state vector differential equation

$$\dot{x} = A \cdot x + B \cdot u$$

(11)

Taking Laplace transforms of (11), we get

$$sX(s) - x(0) = A \cdot X(s) + B \cdot U(s)$$

(12)

or equivalently

$$(s \cdot I\text{-}A) \cdot X(s) = x(0) + B \cdot U(s) \qquad (13)$$

Using a state variable representation of a system, the characteristic equation is given by

$$|(s \cdot I\text{-}A)| = 0 \qquad (14)$$

This yields the characteristics (closed-loop form) equation [16]:

$$a_n \cdot s^n + a_{n-1} \cdot s^{n-1} + \cdots + a_1 \cdot s + a_0 = 0 \qquad 15)$$

The general form of the above system can be expressed in the form of system transfer function as

$$\frac{C}{R}(s) = \frac{G(s)}{1 + G(s) \cdot H(s)} = \frac{K_c \cdot (s - z_{c1}) \cdot (s - z_{c2}) \ldots (s - z_{cn})}{(s - p_{c1}) \cdot (s - p_{c2}) \ldots (s - p_{cn})} \qquad (16)$$

where $s = p_{c1}, p_{c2}, \ldots, p_{cn}$ are closed-loop fuzzy poles, since their values make (16) infinite (also the roots of the characteristic equation) and $s = z_{c1}, z_{c2}, \ldots, z_{cn}$ are closed-loop fuzzy zeros, since their corresponding values of (13) are zero.

We present now the handling of the general form of fuzzy control system modeling and analysis using representative examples of the fourth-order systems.

METHODOLOGY IMPLEMENTATION TO CONTROL SYSTEMS FUZZY IMPULSE RESPONSE

We demonstrate in this section how a fourth order system of the transfer function as expressed by (16) can be handled in a fully fuzzy environment where all the system coefficients are expressed in the Arithmetic fuzzy logic-based representation form. Let us introduce this example that describes the fuzzy response of a high-order control system operating in fully fuzzy environment. We introduce the example in a general form of fourth-order open-loop transfer function, as follows [16]:

$$X_0(s) = \frac{a_0}{s \cdot (s + b_1) \cdot (s^2 + c_1 \cdot s + c_2)} \tag{17}$$

where a_0, b_1, c_1 and c_2 are fuzzy parameters.

Eq. (17) may be written using partial fraction representation as

$$X_0(s) = \frac{A}{s} + \frac{B}{s + b_1} + \frac{C \cdot s + D}{(s + c_1/2)^2 + c_3^2} \tag{18}$$

where $c_3 = \sqrt{c_2 - c_1^2/4}, A, B, C, D$, and c_3 are fuzzy coefficients. Equating coefficient of (17), we get

$$
\begin{aligned}
(s^3): &\quad 0 = A + B + C \\
(s^2): &\quad 0 = A \cdot (b_1 + c_1) + B \cdot c_1 + C \cdot b_1 \\
(s^1): &\quad 0 = A \cdot (c_2 + b_1 \cdot c_1) + B \cdot c_2 + D \cdot b_1 \\
(s^0): &\quad a_0 = A \cdot b_1 \cdot c_2
\end{aligned} \tag{19}
$$

Using the Gaussian Elimination technique, the matrix equation of (19) can be solved with its corresponding fuzzy levels.

Illustrative example 1.

As a numerical example, we choose the value of fuzzy parameters as shown in Table 3. The results of parameters A,B,C, andD are also shown in the same table. Accordingly, the inverse Laplace transform of (19) can be expressed as

$$x_0(t) = A - B \cdot e_{b1} \cdot t - C \cdot e_{c1} \cdot t/2 [(c_4)^2 \cdot \sin_{c3} t - \cos_{c3} t] \tag{20}$$

such that $C_4 = \left(\frac{D}{C} - \frac{c_1}{2}\right)$ where $A,B,C,D,b1,c1,c3,$ and c_4 are fuzzy parameters.

Table 3. Consolidity analysis of the fuzzy impulse response of Illustrative example 1.

Type	Parameter	Value	Fuzzy level scenarios								
			I	II	III	IV	V	VI	VII	VIII	IX
Input	a_0	12.5	3	1	−1	−3	1	4	4	3	5
	b1	0.5	−3	−1	1	3	6	4	5	6	4
	c1	1	−2	−2	2	2	4	4	−4	2	4
	c2	25	3	1	−1	−3	3	6	3	5	6
Output	A	1	3	1	−1	−3	−8	−6	−4	−8	−5
put	B	−1.010	3	1	−1	−3	−8	−6	−4	−8	−5
	C	0.01	−4	−4	4	4	−3	−4	−15	−9	−3
	D	−0.495	−6	0	0	6	−2	−2	1	−2	−1
Consolidity value $FO/(I+S)$[a]	index		0.0019	0.0802	0.0802	0.0185	0.8298	0.3825	0.4096	0.4444	0.1823

[a]Average value of consolidity index
FO/(I+S)=0.2718.

The consolidity pattern of the problem described by plotting the overall output fuzziness factor $|FO|$ versus input fuzziness factor $|FI+S|$ is shown in Fig. 5. The impulse response output solution pattern reveals slight unconsolidated distribution of the results, indicating relatively *low susceptibility* of the optimal solution for change versus any system and input parameters changes effect. Based on consolidity chart of Fig. 5, it can be seen that the control system is almost *consolidated* of class "C".

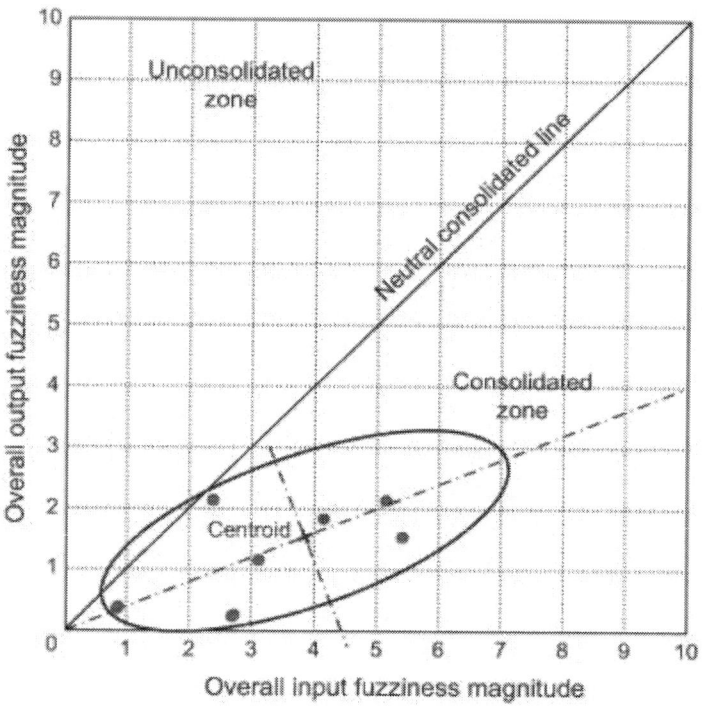

Figure. 5. Consolidity chart of the impulse response problem of Illustrative example 1.

For the selected first four scenarios shown in Table 3, the fuzzy levels of impulse responses are given also in Table 3 and Fig. 6. The equations were solved in Excel sheet with built-in functions programmed using Visual Basic Applications (VBAs). In the implementation procedure, the exact values of fuzzy levels are preserved all over the calculations and are rounded to integer values only at the final result. It follows from the sketches of the impulse time response of Fig. 6 and Table 4 that the fuzziness is related to the time instant. The color of the response is an indication of the fuzzy level using the color coding shown in Table 5. Such colors are selected arbitrarily without restricting that corresponding positive and negative colors are conjugates (summation is either white or black). This is equivalent to the Visual fuzzy logic-based representation [24], [25], [26] and [27].

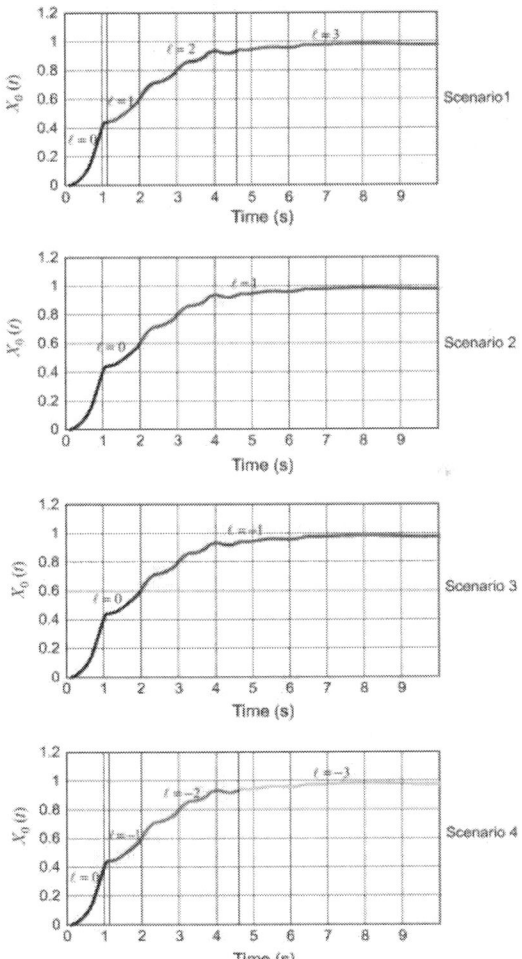

Figure 6. Fuzzy impulse response using visual representation of Illustrative example 1.

Table 4. System impulse response and corresponding fuzzy levels of Illustrative example 1.

Ser.	Time (s)	$x_0(t)$	Selected fuzzy levels scenarios			
			I	II	III	IV
1	0	0	0	0	0	0
2	1	0.3882	1	0	0	−1
3	2	0.6257	1	0	0	−1
4	3	0.7711	2	1	−1	−2
5	4	0.86	2	1	−1	−2
6	5	0.9144	3	1	−1	−3
7	6	0.9477	3	1	−1	−3
8	7	0.968	3	1	−1	−3
9	8	0.9805	3	1	−1	−3
10	9	0.9881	3	1	−1	−3
11	10	0.9927	3	1	−1	−3

Table 5. Definition of positive and negative color sample scales used in sketching Fig. 6.

Ser.	Color	Color code	RBG color index	Excel color index	Corresponding fuzzy level	Type
1		Violet (lavender)	(255, 0, 255)	7	6	
2		Blue	(0, 102, 204)	5	5	
3		Green	(51, 153, 102)	50	4	
4		Violet (lavender light)	(204, 153, 255)	39	3	Positive colors
5		Blue light	(153, 204, 255)	37	2	
6		Green light	(204, 255, 204)	35		

1

7		Black	(0, 0, 0)	2	0	
8		Yellow light	(255, 255, 204)	19	−1	
9		Orange light	(255, 153, 0)	45	−2	
10		Red light	(255, 153, 204)	38	−3	Negative colors
11		Yellow	(255, 255, 0)	27	−4	
12		Orange	(255, 102, 0)	46	−5	
13		Red	(255, 0, 0)	3	−6	

The analysis of the *consolidity chart* of the impulse response problem of Illustrative example 1 shown in Fig. 5 can be summarized as follows:

Symbol	Meaning	Results
R	Shape type	Elliptical
S	Slope	15.21°, or $\tan^{-1}(0.27180)$
$C=(x,y)$	Centroid	(3.8481, 1.6203)
A	Area of shape	13.7414 pu²
l_1	Length of major diagonal	6.9873 pu
l_2	Length of minor diagonal	2.5316 pu
l_2/l_1	Diversity ratio	0.3623

The results indicate that the consolidity region has a very low overall consolidity index and moderate diversity ratio. The area of the consolidity region R is also moderate supporting the moderate diversity of calculated consolidity points.

Similar approaches can be applied for the fuzzy pulse response of digital or discrete data control systems expressed by their z-transform transfer functions [16] and [17].

METHODOLOGY IMPLEMENTATION TO FUZZY ROUTH–HURWITZ STABILITY CRITERION

The work of Routh and Hurwitz [16] gives a method of indicating the presence and number of unstable roots, but not their value. Consider the general form of characteristic equation expressed by (14). The Routh–Hurwitz stability criterion states that for there to be no roots with positive real parts it is a necessary, but not sufficient, condition that all coefficients in the characteristic equation have the same sign and that none is zero. If the above is satisfied, then the necessary and sufficient condition for stability is either

a) All the Hurwitz determinants of the polynomial are positive, or alternatively.
b) All coefficients of the first column of Routh array have the same sign. The number of sign changes indicates the number of unstable roots.

The Hurwitz determinants of (15) can be expressed as follows [16]:

$$D_1 = a_1 \qquad D_2 = \begin{vmatrix} a_1 & a_3 \\ a_0 & a_2 \end{vmatrix}$$

$$D_3 = \begin{vmatrix} a_1 & a_3 & a_5 \\ a_0 & a_2 & a_4 \\ 0 & a_1 & a_3 \end{vmatrix} \qquad D_4 = \begin{vmatrix} a_1 & a_3 & a_5 & a_7 \\ a_0 & a_2 & a_4 & a_6 \\ 0 & a_1 & a_3 & a_5 \\ 0 & 0 & a_2 & a_4 \end{vmatrix} \qquad (21)$$

..., etc., such that all parameters are expressed in the Arithmetic fuzzy logic-based representation form. All the above determinant operations are carried out following the fuzzy logic-based algebra operations [29].

Illustrative example 2.
Let us check the stability of the closed-loop control system of Illustrative example 2, where the open-loop transfer function is expressed in (17), and is having a unity feedback. The closed-loop characteristic function can be expressed as

$$s \cdot (s + b_1) \cdot (s^2 + c_1 s + c_2) + a_0 = 0 \tag{22}$$

or equivalently

$$s^4 + (c_1 + b_1) \cdot s^3 + (c_2 + c_1 b_1) \cdot s^2 + c_2 b_1 \cdot s + a_0 = 0 \tag{23}$$

where a_0, b_1, c_1 and c_2 are fuzzy parameters following the scenarios given in Table 6. Applying now the Routh–Hurwitz criterion, we have first to test that all the coefficients are present and have the same sign. The second test is to check the fuzzy determinants D_1, D_2, and D_3 for different scenarios.

Table 6. Results of Routh–Hurwitz fuzzy determinants of Illustrative example 2.

Ser.	Aspect	Value	Fuzzy level scenarios								
			I	II	III	IV	V	VI	VII	VIII	IX
1	a_0	12.5	1	3	2	4	4	5	3	−4	−1
2	b_1	0.5	3	1	4	1	2	1	−6	−3	−3
3	c_1	1	1	3	3	3	2	2	1	−1	−2
4	c_2	25	1	2	1	2	3	3	3	5	−2
5	$\|D_1\|$	12.5	4	3	5	3	5	4	3	2	−5
6	$\|D_2\|$	300	5	5	6	5	8	7	0	8	−7
7	$\|D_3\|$	443.75	6	8	8	8	10	9	9	7	−9
Consolidity index$_{FO/(I+S)}$[a]			5.7855	2.8442	5.6727	2.4726	2.8328	2.2539	1.9153	3.7299	4.9909

[a]Average value of consolidity index
FO/(I+S)=3.6149.

The Hurwitz determinants for this numerical example can be expressed as

$$D_1 = c_2 \; b_1 \tag{24}$$

$$D_2 = \begin{vmatrix} c_2 \cdot b_1 & c_1 + b_1 \\ a_1 & c_2 + c_1 \cdot b_1 \end{vmatrix} \tag{25}$$

And

$$D_3 = \begin{vmatrix} c_2 \cdot b_1 & c_1 + b_1 & 0 \\ a_1 & c_2 + c_1 \cdot b_1 & 1 \\ 0 & c_2 \cdot b_1 & c_1 + b_1 \end{vmatrix} \tag{26}$$

The numerical results are shown for example in Table 6.

The Hurwitz determinants of the polynomial are all positive with various level of fuzziness. The fuzzy levels of determinants describe the level of uncertainty in the results. For different scenarios, we will have same fuzzy results with opposite signs of the fuzziness of the determinant of scenarios corresponding with opposite input fuzzy data. This indicates the inverse property of the Arithmetic fuzzy logic-based approach. Similar approach can be applied for solving Jury stability criterion of fuzzy discrete control systems expressed by the z-transform characteristic equation of the sampled data system [16] and [17].

The consolidity pattern of the problem described by plotting the overall output fuzziness factor $|_{FO}|$ versus input fuzziness factor $|_{FI+S}|$ is shown in Fig. 7. The solution pattern reveals high unconsolidated distribution of the results, indicating *high susceptibility* of the Routh–Hurwitz stability criterion for change versus any system and input parameters changes effect. Based on consolidity chart of Fig. 7, it can be seen that the control system is *unconsolidated* of class "*U*".

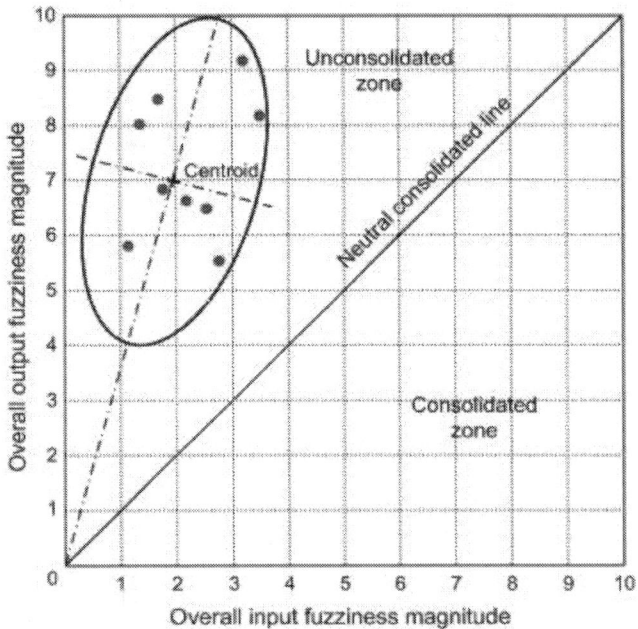

Figure 7. Consolidity chart of the Routh–Hurwitz stability criterion of Illustrative example 2.

The analysis of the consolidity chart of the Routh–Hurwitz stability criterion of Illustrative example 2 sketched in Fig. 7 can be summarized as follows:

Symbol	Meaning	Results
R	Shape type	Elliptical
S	Slope	74.54°, or $\tan^{-1}$ (3.6149)
$C=(x,y)$	Centroid	(1.9241, 7.0886)
A	Area of shape	14.7669 pu²
l_1	Length of major diagonal	6.1772 pu
l_2	Length of minor diagonal	3.0379 pu
l_2/l_1	Diversity ratio	0.4918

The results reveal that the consolidity region has both moderate overall consolidity index and diversity ratio. The area of the consolidity region R is moderate supporting the moderate diversity of calculated consolidity.

METHODOLOGY IMPLEMENTATION TO CONTROL SYSTEMS FUZZY CONTROLLABILITY AND OBSERVABILITY

A system is said to be *controllable* if a control vector u(t) exists that will transfer the system from any initial state $x(t_0)$ to some final state x(t) in a finite time interval.

A system is said to be *observable* if at time t_0, the system state $x(t_0)$ can be exactly determined from observation of the output y(t) over a finite time interval.

If the system is described by

$$\dot{x} = A \cdot x + B \cdot u$$
$$y = C \cdot x + D \cdot u \tag{27}$$

then a sufficient condition for complete state controllability is that the $n \times n$ matrix [16]:

$$M = [B : A \cdot B : \ldots : A^{n-1} \cdot B] \tag{28}$$

contains n linearly independent row or column vectors, i.e. is of rank n (that is, the matrix is non-singular, and the determinant is non-zero). Eq. (28) designates the*Controllability matrix.*

The system described by (28) is completely observable if the $n \times n$ matrix is denoted as the **Observability matrix** [16]:

$$N=[C^T \vdots A^T \cdot C^T \vdots \cdots \vdots (A^T)^{n-1} \cdot C^T] \tag{29}$$

where all the system coefficients are expressed in the Arithmetic fuzzy logic-based representation form.

Illustrative example 3.
Consider the state space representation of the closed-loop system of Illustrative example 3:

$$\begin{bmatrix} x_1^\bullet \\ x_2^\bullet \\ x_3^\bullet \\ x_4^\bullet \end{bmatrix} = \begin{bmatrix} 0 & 1 & 0 & 0 \\ 0 & 0 & 1 & 0 \\ 0 & 0 & 0 & 1 \\ -a_0 & -a_1 & -a_2 & -a_3 \end{bmatrix} \begin{bmatrix} x_1 \\ x_2 \\ x_3 \\ x_4 \end{bmatrix} + \begin{bmatrix} 0 \\ 0 \\ 0 \\ \alpha \end{bmatrix} u \tag{30}$$

And

$$y = [0 \quad 0 \quad 0 \quad \beta] \cdot \begin{bmatrix} x_1 \\ x_2 \\ x_3 \\ x_4 \end{bmatrix} \tag{31}$$

such that $a_1 =_{c2} b_1, a_2 =_{c2} +_{c1} b_1, a_3 =_{c1} +_{b1}$ and α, β are fuzzy parameters, where $\alpha = 2$ and $\beta = 1$.

Let us form the Controllability matrix M as

$$M = [B : A \cdot B : A^2 \cdot B : A^3 \cdot B]$$

$$= \alpha^4 \begin{bmatrix} 0 & 0 & 0 & 1 \\ 0 & 0 & 1 & -a_3 \\ 0 & 1 & -a_3 & -a_2 + a_3^2 \\ 1 & -a_3 & -a_2 + a_3^2 & -a_1 + 2 \cdot a_2 \cdot a_3 - a_3^3 \end{bmatrix}$$

(32)

This yields the determinant of $M = \alpha^4 \neq 0$, with fuzzy level = 4 ι_α. Thus the system is controllable. The fuzzy levels of various scenarios are illustrated in Table 7. The results indicate that for this example the fuzzy level of the controllability matrix determinant is a function only of α^4 and not related to other system parameters fuzziness.

Table 7. Consolidity results controllability and observability matrix criteria of Illustrative example 3.

Ser.	Aspect	Value	Fuzzy level scenarios								
			I	II	III	IV	V	VI	VII	VIII	IX
1	a0	12.5	−1	6	3	5	8	5	3	6	3
2	a1	12.5	2	5	7	4	7	3	9	3	10
3	a2	25	3	2	4	2	3	2	5	1	5
4	a3	1.5	−2	4	2	2	3	4	2	3	2
5	α	2	−1	−2	2	2	2	1	2	−1	2
6	β	1	2	−4	1	−3	−4	−1	1	−1	1
7	$\|M\|$	16	−4	−8	8	8	8	4	8	−4	8
Consolidity value$_{FO/(I+S)}$[a]		index	1.977	3.953	2.777	3.672	2.358	2.061	2.309	2.605	2.301
8	$\|N\|$	−58.085 × 10³	9	−8	9	−6	−7	4	8	3	8
Consolidity value$_{FO/(I+S)}$[b]		index	4.552	4.123	3.059	2.746	1.927	1.892	2.419	2.087	2.375

[a]Average value of consolidity index FO/(I+S)=2.6681.
[b]Average value of consolidity index FO/(I+S)=2.7976.

Applying the Observability Matrix criterion, we have

$$N = \left[C^T : A^T \cdot C^T : \left(A^T\right)^2 \cdot C^T : \left(A^T\right)^3 \cdot C^T \right]$$

$$= \beta^4 \begin{bmatrix} 0 & -a_0 & a_0 a_3 & a_0 \cdot a_2 - a_0 \cdot a_3^2 \\ 0 & -a_1 & -a_0 + a_1 \cdot a_3 & a_0 \cdot a_3 + a_1 \cdot a_2 - a_1 \cdot a_3^2 \\ 0 & -a_2 & -a_1 + a_2 \cdot a_3 & -a_0 + a_1 \cdot a_3 + a_2^2 - a_2 \cdot a_3^2 \\ 1 & -a_3 & -a_2 + a_3^2 & -a_1 + 2a_2 \cdot a_3 - a_3^3 \end{bmatrix}$$

$$\tag{33}$$

The determinant of $N \neq 0$, thus the system is also observable. The associated fuzzy levels of various scenarios are also given in Table 7. It follows from the example that the fuzzy level of the Observability matrix determinant is a function of β^4 and also related to the system parameter fuzziness. The results indicate the existence of a relatively high fuzziness of the Observability Matrix due to the fuzziness of the system's parameters.

The consolidity pattern of the controllability and observability problem of Illustrative example 3 is described by plotting the output fuzziness factor $|\text{FO}|$ versus input fuzziness factor $|\text{FI+S}|$ as shown in Fig. 8 and Fig. 9. Both charts reveal moderately unconsolidated distribution of the results, indicating relatively *medium susceptibility* of the both conditions for change versus any system and input parameters changes effect. Based on consolidity charts of Fig. 8 and Fig. 9, it can be seen that the control system performance is *unconsolidated* of class "U" for both controllability and observability.

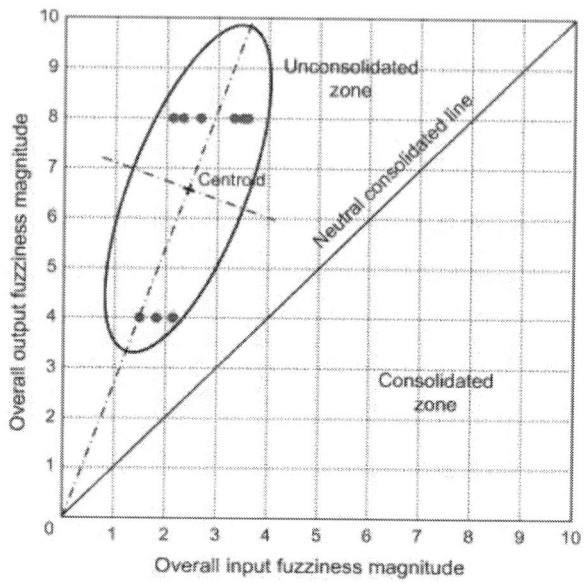

Figure 8. Consolidity pattern of the fuzzy system *controllability* results of Illustrative example 3.

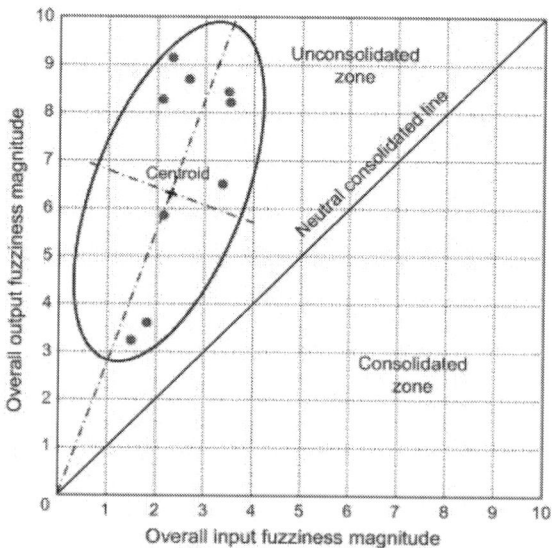

Figure 9. Consolidity pattern of the fuzzy system *observability* results of Illustrative example 3.

The analysis of the consolidity chart of the fuzzy system *controllability* results ofIllustrative example 3 sketched in Fig. 8 can be summarized as follows:

Symbol	Meaning	Results
R	Shape type	Elliptical
S	Slope	69.45° or $\tan^{-1}$ (2.6681)
$C=(x,y)$	Centroid	(2.3291,6.3797)
A	Area of shape	18.9713 pu^2
l_1	Length of major diagonal	7.5949 pu
l_2	Length of minor diagonal	3.1392 pu
l_2/l_1	Diversity ratio	0.4133

The results show that the consolidity region has a moderate overall consolidity index and relatively high diversity ratio. The area of the consolidity region is very high supporting the high diversity of calculated consolidity points.

As for the consolidity chart of the fuzzy system *observability* results of Illustrative example 3 delineated in Fig. 9, the analysis can be summarized as follows:

Symbol	Meaning	Results
R	Shape type	Elliptical
S	Slope	70.33° or $\tan^{-1}$ (2.7976)
$C=(x,y)$	Centroid	(2.4304,6.5822)
A	Area of shape	13.7414 pu^2
l_1	Length of major diagonal	6.9873 pu
l_2	Length of minor diagonal	2.3291 pu
l_2/l_1	Diversity ratio	0.3333

The results show that the consolidity region has a moderate overall consolidity index and low diversity ratio. The area of the consolidity region R is high supporting the spread of calculated consolidity points.

Similar approach can be applied for testing the fuzzy controllability and observability of digital control systems expressed by their discrete time state equations [16] and [17].

METHODOLOGY IMPLEMENTATION TO MODELING AND DESIGN OF INVERTED PENDULUM STABILIZATION

In this section, the suggested approach is implemented for the fuzzy modeling and stabilization of the inverted pendulum system (to be referred to as **Illustrative example 4**) as shown in Fig. 10. The inverted pendulum problem is an example of producing a stable closed-loop control system from an unstable plant. For this system, it is possible to design a controller using the pole placement techniques [16].

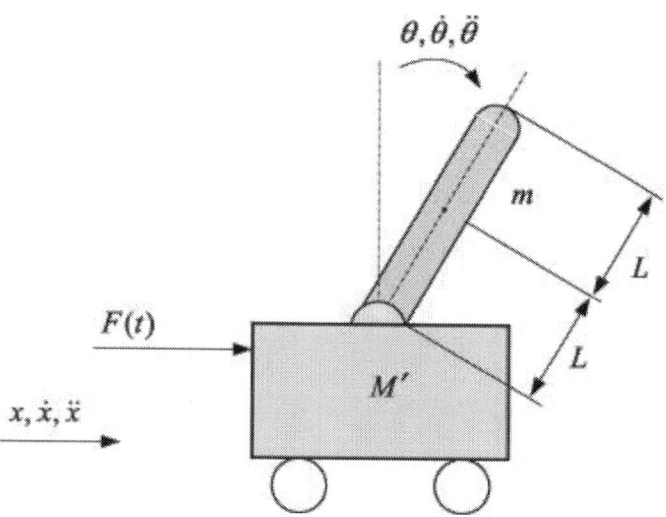

Figure 10. Sketch showing main parameters of the inverted pendulum of **Illustrative example 4**.

In the figure, m is the mass of pendulum, L denotes the half-length of the pendulum and M' is the mass of the trolley. The parameter $F(t)$ indicates the applied force to the trolley in the x-direction.

It is assumed that θ is small and the second-order terms $(\dot{\theta}^2)$ can be neglected, then we can define the state variables of the inverted pendulum system as $(g=9.81)$

$$x_1 = \theta, \quad x_2 = \dot{\theta}, \quad x_3 = x \quad \text{and} \quad x_4 = \dot{x} \tag{34}$$

and the control variable is

$$u=F(t) \tag{35}$$

From (34) and (35), the state equations become

$$
\begin{bmatrix} \dot{x}_1 \\ \dot{x}_2 \\ \dot{x}_3 \\ \dot{x}_4 \end{bmatrix} = \begin{bmatrix} 0 & 1 & 0 & 0 \\ a_{21} & 0 & 0 & 0 \\ 0 & 0 & 0 & 1 \\ a_{41} & 0 & 0 & 0 \end{bmatrix} \begin{bmatrix} x_1 \\ x_2 \\ x_3 \\ x_4 \end{bmatrix} + \begin{bmatrix} 0 \\ b_2 \\ 0 \\ b_4 \end{bmatrix} u \tag{36}
$$

Where

$$a_{21} = \frac{3 \cdot g \cdot (M' + m)}{L \cdot [4 \cdot (M' + m) - 3 \cdot m]}$$

$$a_{41} = \frac{-3g \cdot m}{4 \cdot (M' + m) - 3 \cdot m}$$

$$b_2 = \frac{-3}{L \cdot [4 \cdot (M' + m) - 3 \cdot m]}$$

$$b_4 = \left(\frac{1}{M' + m}\right) \cdot \left\{1 + \frac{3 \cdot m}{4 \cdot (M' + m) - 3 \cdot m}\right\} \tag{37}$$

$$
M = \begin{bmatrix} 0 & b_2 & 0 & a_{21} \cdot b_2 \\ b_2 & 0 & a_{21} \cdot b_2 & 0 \\ 0 & b_4 & 0 & a_{41} \cdot b_2 \\ b_4 & 0 & a_{41} \cdot b_2 & 0 \end{bmatrix}
$$
(38)

Data for simulation are represented in Table 8 for different selected scenarios. The output equation is

$$y = C\,x$$
(39)

where C is the identity matrix. For a regulator, with a scalar control variable and gain vector K, we have

$$u = -K\,x$$
(40)

The elements of K can be obtained by selecting a set of desired closed-loop poles by the Ackermann's formula [16] for system stabilization through the pole placement technique. Let

$$K = \begin{bmatrix} 0 & 0 & 0 & \cdots & 1 \end{bmatrix} \cdot M^{-1} \cdot \phi(A)$$
(41)

where M is the **Controllability matrix** and

$$\phi(A) = A^n + a_{n-1} \cdot A^{n-1} + \cdots + a_1 \cdot A + a_0 \cdot I$$
(42)

where A is the system matrix and a_i represent the coefficients of the desired closed-loop characteristic equation.

If the required closed-loop poles are $s=-2\pm j2$ for the pendulum, and $s=-4\pm j4$ for the trolley, then the closed-loop characteristic equation becomes

$$s^4+12s^3+72s^2+192s+256=0 \tag{43}$$

The algebraic form of the fuzzy gain result can be expressed as follows: let

$$[\beta_1 \quad \beta_2 \quad \beta_3 \quad \beta_4] = [0 \quad 0 \quad 0 \quad 1]M^{-1} \tag{44}$$

then we can attain by simple matrix operation the following fuzzy values of the gain vector K:

$$k_1 = \beta_1 \cdot (\alpha_o + \alpha_2 \cdot a_{21} + \alpha_4 \cdot a_{21}^2) + \beta_2(\alpha_2 \cdot a_{41} + \alpha_4 \cdot a_{21} \cdot a_{41}) \tag{45}$$

$$k_2 = \beta_1 \cdot (\alpha_1 + \alpha_3 \cdot a_{21}) + \beta_3 \cdot \alpha_3 \cdot a_{41} \tag{46}$$

$$k_3 = \beta_3 \cdot \alpha_0 \tag{47}$$

and

$$k_4 = \beta_3 \cdot \alpha_1 \tag{48}$$

For design purposes, various prototypes of the inverted pendulum can be selected with different relative trolley and pendulum masses of different manufacturing materials and with gain vector K satisfying the same stabilized characteristics function of (34). The most desired robust design can be attained that achieves the best consolidity performance. In this respect, five different designs of trolley car masses M' of 8, 4, 2, 1 and 0.6 respectively are selected as shown in Fig. 11.

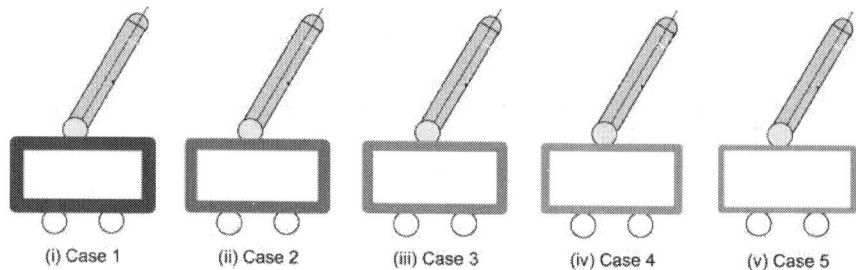

(i) Case 1   (ii) Case 2   (iii) Case 3   (iv) Case 4   (v) Case 5

Figure 11. Various selected design prototypes of inverted pendulum of different manufacturing materials of **Illustrative example 4** (Case 1: M' = 8, Case 2: M' = 4, Case 3: M' = 2, Case 4: M' = 1, Case 5: M' = 0.6).

For each design, the consolidity analysis is applied and the result is shown in Table 8 for Case 1 of M' = 8 and sketched all the three cases in Fig. 12. From this figure, it appears that the consolidity pattern of the inverted pendulum improves with the reduction of the trolley car mass M' and the best consolidity performance is obtained for the fifth design case with M' = 0.6. The analysis of the consolidity chart of the fuzzy inverted pendulum of various selected designs of **illustrative example 4** delineated in Fig. 12 can be summarized as follows:

Symbol	Results of various designs of M'				
	8	4	2	1	0.6
R	Elliptical	Elliptical	Elliptical	Elliptical	Elliptical
S	74.51° or $\tan^{-1}(3.6080)$	63.96° or $\tan^{-1}(2.0467)$	51.42° or $\tan^{-1}(1.2534)$	39.60° or $\tan^{-1}(0.8274)$	32.23° or $\tan^{-1}(0.6305)$
$C=(x,y)$	(0.5250, 1.9250)	(0.8500, 1.8400)	(1.4500, 1.8450)	(2.0750, 1.5000)	(2.5000, 1.5000)
$A(pu^2)$	1.25	1.275	1.65	2.275	2.5
$l_1(pu)$	1.95	1.95	2.2	2.4	2.55
l_2	0.8	0.85	0.95	1.2	1.25
l_2/l_1	0.4103	0.4359	0.4318	0.5	0.4902

Table 8. Consolidity analysis of inverted pendulum of Case 1: M' = 8 of **Illustrative example 4**.

Type	Parameter	Value	Corresponding fuzzy level values								
			I	II	III	IV	V	VI	VII	VIII	IX
Input	L	1	2	2	4	3	2	3	4	3	4
	m	0.5	3	4	3	5	2	6	5	3	7
	a_{21}	9.81	−2	−2	−4	−3	−2	−3	−4	−3	−4
	a_{41}	−3.27	−3	−4	−4	−5	−2	−6	−5	−3	−7
	b_2	−0.6667	−2	−2	−4	−3	−2	−3	−4	−3	−4
	b_4	0.8889	0	0	0	0	0	0	0	0	0
Output	$|M|$	0.007	−8	−8	−16	−11	−8	−11	−15	−12	−15
	k_1	−212.0	1	1	3	2	1	2	3	2	3
	k_2	282.7	1	0	1	1	1	1	1	1	1
	k_3	−412.7	2	2	4	3	2	3	4	3	3
	k_4	−1240.3	2	2	4	3	2	3	4	3	3
Consolidity index value$_{FO/(I+S)}$[a]			3.4779	3.022	4.4727	3.3121	4.0858	3.022	3.7585	4.0858	3.2348

[a]Average value of consolidity index FO/(I+S)=3.6080.

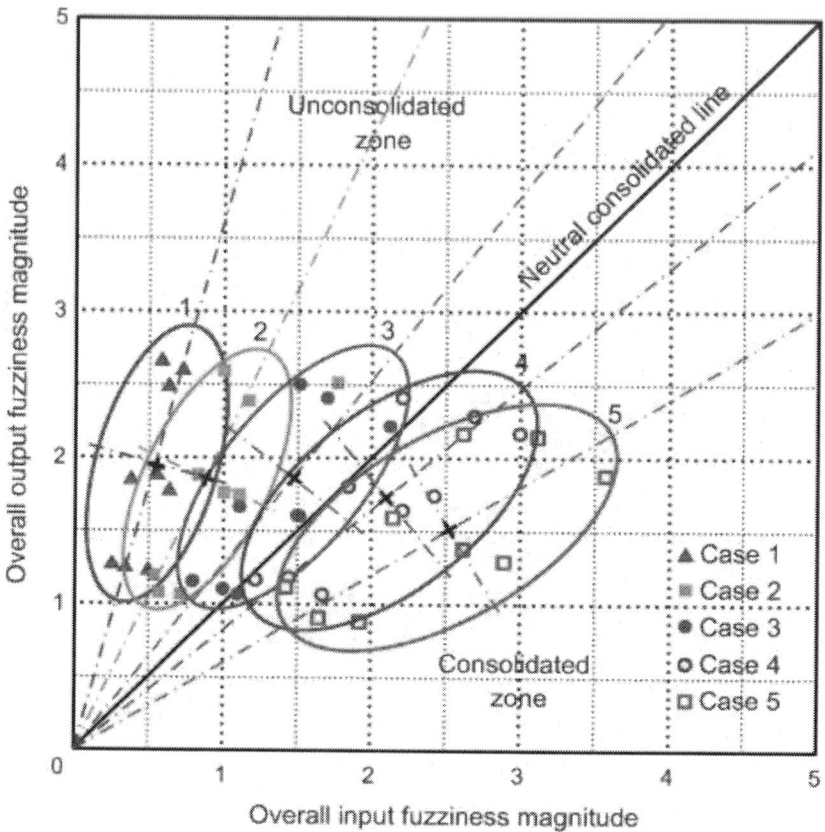

Figure 12. Consolidity patterns of the fuzzy inverted pendulum designs of **Illustrative example 4**.

It follows from the above table that reducing the trolley weight the system overall consolidity is improved but is accompanied with increase of the consolidity region areas. In general, the diversity ratio is high indicating high diversity of calculated consolidity points. The areas of the regions are in general very small compared to all previous illustrative examples ($A > 13$). Such charts of Fig. 12 clearly demonstrate the effectiveness of the proposed approach as a tool for the analysis and design of automatic control systems.

CONCLUSIONS

A new approach for the fuzzy automatic control systems' modeling and analysis using the consolidity theory was presented. Key implementations issues in fuzzy automatic control and dynamical systems were addressed in a very smooth and systematic way. These issues covered system's fuzzy impulse response, system's stability using Routh–Hurwitz criterion, system's Controllability and Observability, and the stabilization of inverted pendulum through the pole placement technique. Illustrative examples of fourth-order systems were solved to demonstrate the effectiveness and applicability of the new technique. The approach is also suitable for higher dimensional automatic control and dynamical systems as it is based mainly on matrix formulations.

Implementation procedures were elaborated for the consolidity analysis of existing control systems and the design of new ones. Systems comparisons based on such implementation consolidity results were discussed based on the values of the average, weighted or worst consolidity scores, or the comparison of the plotted regions (patterns) of fuzziness behavior in the sketched consolidity chart. It is shown that the regions comparison in consolidity chart is based on type of consolidity region shape, slope or angle in degrees of the centerline of the geometric shape, the centroid of the geometric shape, area of the geometric shape, length of principal diagonals of the shape, and the diversity ratio of consolidity points for each region.

The suggested approach could open the door for more general modeling, analysis and design of both *continuous* and *digital* data automatic control systems. Moreover, it provides an effective tool for designing new control systems that could withstand future changes due to any system or input parameters changes effects on and above normal systems operations or set points. Examples of some foreseen extensions are as follows:

(i) Design of Fuzzy P, PI, PD, and PID control systems.
(ii) State-space methods for fuzzy automatic control system analysis and design.

(iii) Design of fuzzy state observers and estimators for closed-loop systems.

(iv) Optimal and robust fuzzy control of multivariate systems.

(v) Other problems such as multivariate fuzzy Kalman state estimation, fuzzy linear quadratic regulators, and fuzzy Lyapunov stability criterion.

It was shown that the presented consolidity methodology is open in its application to wide classes of systems. Even for the system that thought not to be fuzzy, we can still imagine that these systems are operating in a fully fuzzy environment and perform typically the same consolidity testing. Its needless to say that all the present physical systems in our daily life are subject to continuous wearing and deterioration that make them gradually changing, and thus will behave later equivalently as if they are operating in a fuzzy environment. This makes the presented *consolidity chart* approach *generally* extendable to wider spectrum of real-life applications beside that given in the automatic control fields. Examples of these fields are geology, archeology, life sciences, ecology, environmental science, engineering, materials, medicine, biology, sociology, humanities, and many other important fields.

ACKNOWLEDGMENTS

The author is indebted to the reviewers for their constructive comments that have led to improving considerably the analysis given in the paper.

REFERENCES

1. Feng E. A survey on analysis and design of model-based fuzzy control systems. IEEE Trans Fuzzy Syst 2006;14(5):676–97.

2. El-Kholy EE, Dab Room AM, El-Khloy AE. Adaptive fuzzy logic controllers for dc drives: a survey of the state of art. J Electr Syst 2006;2–3:116–45.

3. Boulkroune A, Tadjine M, Saad MM, Farza M. Fuzzy adaptive controller for MIMO nonlinear systems with known and unknown control direction. Fuzzy Sets Syst 2010;161:797–820 Elsevier..

4. Gao S, Zaiyue Z, Cao C. Multiplications operation on fuzzy numbers. J Softw 2009;4(4):331–8.

5. Yan F, Lu S. On-line inference for fuzzy controllers in continuous domains. In: Fuzzy information and engineering. AISc 62, vol. 2. Berlin Heidelberg: Springer-Verlag; 2009. p. 1111–8.

6. Ramle N, Mohamed D. A centroid-based performance evaluation using aggregated fuzzy numbers. Appl Math Funct 2009;3(48): 2369–81.

7. Garibaldi JM, Jaroszewski M, Musikasuwan S. Non-stationary fuzzy sets. IEEE Trans Fuzzy Syst 2008;16(4):1072–886.

8. Ramli N, Mohamed D. A comparative analysis of centroid methods in ranking fuzzy numbers. Eur J Sci Res 2009;28(3): 492–501.

9. Kubica E, Madill D, Wang D. Designing stable MIMO fuzzy controllers. IEEE Trans Syst Man Cybernet – Part B: Cybernet 2005;35(2):372–80.

10. Tanaka K, Wang HO. Fuzzy control systems design and analysis. New York: John Wiley & Sons, Inc.; 2001.

11. Lam HK, Leung FHF, Tam PKS. Stable and robust fuzzy control for uncertain nonlinear systems. IEEE Trans Syst Man Cybernet – Part A: Syst Hum 2000;30(6):825–40.

12. Mendel JM. Type 2 fuzzy sets and systems: an overview. IEEE Comput Intell Mag 2007:20–9.

13. Azar AH. Fuzzy systems. Vukovar, Croatia: Intech Publishing; 2010.

14. Sbi Y, Eberhart R, Chin Y. Implementation of evolutionary fuzzy systems. IEEE Trans Fuzzy Syst 1999;7(2):109–19.

15. Shaocheng T, Changying L, Yongming L. Fuzzy adaptive observer backstepping control for MIMO nonlinear systems. Fuzzy Sets Syst 2009;160:2755–75 Elsevier..

16. Burns RS. Advanced control engineering. Oxford, UK: Butterworth Heinemann, Reed Elsevier Group; 2001.

17. Chen CT. Analog and digital control system design: transfer functions, state space and algebraic methods. New York: Saunders College Publishing; 1993.

18. Ogata K. Modern control engineering. Upper Saddle River, New Jersey: Pearson Education International, Prentice Hall; 2010.

19. Tsaukalcs LH, Uhrig RE. Fuzzy and neural approaches in engineering. New York: John Wiley & Sons Inc.; 1997.

20. Passino KM, Yukovich S. Fuzzy control. Menlo Park, California, USA: Addison-Wesley Longman, Inc.; 1998.

21. Vasantha Kandasamy WB, Smarandache F. Fuzzy interval matrices, neutrosophic interval matrices and their applications. Phoenix, Arizona: HEXIS; 2006, .

22. Vasantha Kandasamy WB, Smarandache F, Amal K. Super fuzzy matrices and super fuzzy models for social scientists. Ann Arbor, USA: Infollarnquest (ILQ); 2008, .

23. Vasantha Kandasamy WB, Smarandache F, Ilanthenral K. Introduction to bimatrices. Phoenix (Arizona, USA): Hexis; 2005, .

24. Gabr WI, Dorrah HT. New fuzzy logic-based arithmetic and visual representations for systems' modeling and optimization. In: Proceedings of IEEE international conference on robotics and biomimetics (ROBIO 2008), Bangkok, Thailand, February 22–25; 2009. p. 715–22.

25. Gabr WI, Dorrah HT. Development of fuzzy logic-based arithmetic and visual representations for systems' modeling and optimization of interconnected networks. In: Proceedings of IEEE international conference on robotics and biomimetics (ROBIO 2008), Bangkok, Thailand, February 22–25; 2009. p. 723–30.

26. Dorrah HT, Gabr WI. Multi-objective linear optimization using fuzzy logic-based arithmetic and visual representations with forward and backward tracking. In: Proceedings of IEEE international conference on robotics and biomimetics (ROBIO 2008), Bangkok, Thailand, February 22–25; 2009. p. 731–8.

27. Dorrah HT, Gabr WI. Development of fuzzy logic-based arithmetic and visual representations for solving quadratic programming in fully fuzzy environment. In: Proceedings of IEEE international conference on information and automation (ICIA 2009), Zhuhai, China, June 22–25; 2009. p. 46–53.

28. Dorrah HT, Gabr WI. Generalization of arithmetic and visual fuzzy logic-based representations for nonlinear modeling and optimization in fully fuzzy environment. In: Proceedings of IEEE international conference on robotics and biomimetics (ROBIO 2009), Guilin, Guangxi, China, December 18–22; 2009. p. 1013– 20.

29. Gabr WI. Arithmetic fuzzy logic-based representation approach versus conventional fuzzy theory for modeling and analysis in fully fuzzy environment. In: Proceedings of IEEE international conference on information and automation (ICIA 2010), Harbin, Heilongjiang, China, Paper No. 1031751, June 20–23; 2010.

30. Gabr WI. Analogy between arithmetic fuzzy logic-based representation approach and conventional fuzzy theory with applications to operations research. J Al Azhar Univ Eng Sector, ISSN 1110-6409, JAUES 2015; 10(1), Paper No. 35.

31. Dorrah HT, Gabr WI. Foundations of new systems' consolidity theory using arithmetic fuzzy logic-based representation in fully fuzzy environment. In: Proceedings of 6th annual IEEE conference on automation science and engineering (IEEE CASE), Toronto, Ontario, Canada, August 21–24; 2010. p. 624–631. Paper No. MoC2.1.

32. Dorrah HT, Gabr WI. Development of new consolidity theory for systems' analysis and design in fully fuzzy environment. Int J Expert Syst Appl 2011. http://dx.doi.org/10.1016/ i.eswa.2011.07.125 and also ESWA 2012;39(1):1191–9..

33. Dorrah HT. Consolidity: mystery of inner property of systems uncovered. Elsevier J Adv Res 2011. http://dx.doi.org/10.1016/ j.jare.2011.11.002 and also JARE 2012;3(4):345–58..

34. Gabr WI. Consolidity analysis of fuzzy functions, matrices, probability and statistics. Elsevier Ain Shams Eng J 2015. http:// dx.doi.org/10.1016/j.asej.2014.09.014 and also, ASEJ 2015; 6(1):181-197..

35. Dorrah HT. Consolidity: moving opposite to built-as-usual systems practices. Elsevier Ain Shams Eng J 2012. http:// dx.doi.org/10.1016/j.asej.2012.7.004 and also, ASEJ 2013;4(2):221–39..

36. Dorrah HT. Supplement to "Consolidity: moving opposite to built-as-usual systems practice". Elsevier Ain Shams Eng J 2013. http://dx.doi.org/10.1016/j.asej.2013.02.009 and also ASEJ 2013;4(4):783–803..

37. Dorrah HT. Consolidity: stack-based systems change pathway theory elaborated. Elsevier Ain Shams Eng J 2014. http:// dx.doi.org/10.1016/j.asej.2013.12.002 and also ASEJ 2014;5(2):449–73..

38. Dorrah HT. Toward a new "Fractals-general science". Elsevier Alexandria Eng J Short Commun 2014. http://dx.doi.org/10.1016/ j.aej.2014.07.003 and also AEJ 2014;53(3):505–12].

CITATION

Walaa Ibrahim Gabr, A new approach for automatic control modeling, analysis and design in fully fuzzy environment, Ain Shams Engineering Journal, Available online 1 April 2015, ISSN 2090-4479, http://dx.doi.org/10.1016/j.asej.2015.01.010.

CHAPTER 4

Design of an Optimal SMES For Automatic Generation Control of Two-Area Thermal Power System Using Cuckoo Search Algorithm

Sabita Chaine, M. Tripathy

Department of Electrical Engineering, Veer Surendra Sai University of Technology, Burla, Odisha 768018, India

ABSTRACT

This work presents a methodology adopted in order to tune the controller parameters of superconducting magnetic energy storage (SMES) system in the automatic generation control (AGC) of a two-area thermal power system. The gains of integral controllers of AGC loop, proportional controller of SMES loop and gains of the current feedback loop of the inductor in SMES are optimized simultaneously in order to achieve a desired performance. Recently proposed intelligent technique based algorithm known as Cuckoo search algorithm (CSA) is applied for optimization. Sensitivity and robustness of the tuned gains tested at different operating conditions prove the effectiveness of fast acting energy storage devices like SMES in damping out oscillations in power system when their controllers are properly tuned.

INTRODUCTION

It is a well known fact that, any mismatch between generation and load in an interconnected power system causes instability that deteriorates the system dynamic performances disturbing the equilibrium of real power of the system, which in turn affects the system frequency. In this regard, the purpose of AGC is to develop a control system which should be able to maintain the system frequency and tie line real power flowing between different control areas at their respective specified nominal values when the system is subjected to load variations.

With a view to achieve the above objective, one important review work in the field of AGC has tried to comprehensively discuss various control strategies adopted till today (Ibraheem and Kothari, 2005). Effects of the mechanical governor, electric governor, a single stage reheat turbine and a two-stage reheat turbine, on the dynamic responses have been explored by Nanda et al. (2006). Some other works have formulated the problem in the domain of optimization and tried to investigate into the application of both conventional (Mallesham et al., 2010) as well as heuristic optimization techniques (Ghoshal, 2004).

Besides the method of gain tuning through optimization, research works have also tried to examine the efficacies of other power electronics based devices in the family of flexible AC transmission systems (FACTs) (Bhatt et al., 2010), to damp out oscillations in the frequency and tie-line power exchanges. Fast-acting energy storage device such as SMES system has also been found (Banerjee et al., 1990 and Tripathy et al., 1992) to introduce required damping in these oscillations.

This work aims at obtaining an optimal controller for SMES in a two-area power system (Elgerd, 2005) which should exhibit robustness in its performance for a varying operating conditions and parameters of the system. Two most important issues mentioned below, which decide the effectiveness of any such tuned controller, are emphasized while formulating and solving the problem. They are

(i) The nature of designed objective function.

(ii) The efficiency of optimization method.

Objective functions are suitably designed from both time domain and frequency domain perspectives and after optimization their relative performances are tested when subjected to perturbations. In order to optimize the problem, recently proposed nature-inspired metaheuristic algorithms known as Cuckoo search (CS) (Yang and Deb, 2009) have been utilized.

The paper is organized as follows. Section 2 illustrates the system model, and its main components. In Section 3, a brief overview on SMES and its proposed control strategy is presented. Section 4 discusses about the different objective functions which are optimized to maximize the performance of the SMES in AGC domain. A brief overview of the intelligent technique based optimization algorithm CS is elaborated in Section 5. The simulation and results, obtained following several tests related to the performance of tuned SMES, are explained and analyzed in Section 6. At the end, conclusions are presented in Section 7.

AGC IN TWO-AREA THERMAL POWER SYSTEM WITH SMES

Many problems in AGC, particularly related only to the automatic load frequency control (ALFC) part of AGC within two interconnected areas of power system, have utilized a widely accepted model (Elgerd, 2005) in order to examine the response of power system towards several factors including changes in system parameter, model parameters, operating condition, gains of controllers, etc. Fig. 1 depicts the outline of this model, where the blocks of transfer functions representing the governor system, steam reheats turbines, regulation droop R, frequency bias constant, β, etc. are connected. The area control error (ACE) is defined by combining Δf and $\Delta PTie$ as depicted in Eqs. (1) and (2), which are widely used:

$$e_1(t) = \text{ACE}_1 = \beta_1 \Delta f_1 + \Delta P_{Tie} \qquad (1)$$

$$e_2(t) = \text{ACE}_2 = \beta_2 \Delta f_2 - \Delta P_{Tie} \qquad (2)$$

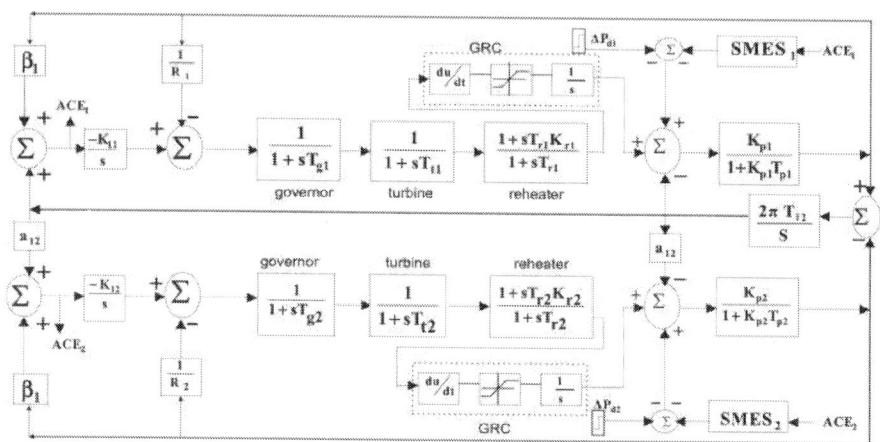

Figure 1. Two-area interconnected power system with SMES unit considering GRC.

The role of SMES in the problem of AGC

The presence of SMES in the control of frequency in an AGC framework provides rapid recovery in the requirement of deficit or surplus real power, by deriving the same from a large inductor or reactor. As per the need of the power system, the power delivered or recovered from the reactor can be controlled by suitably designed controller dedicated for the SMES. A detailed overview behind the fundamental physics and some elementary modelling issues shall be covered in Section 3.

SUPER CONDUCTING MAGNETIC ENERGY STORAGE (SMES) SYSTEM: BRIEF DISCUSSION

As depicted in Fig. 2, the SMES system has a DC magnetic coil that is connected to the AC grid through a power conversion system

(PCS) which includes two numbers of converters for inversion and rectification purposes.

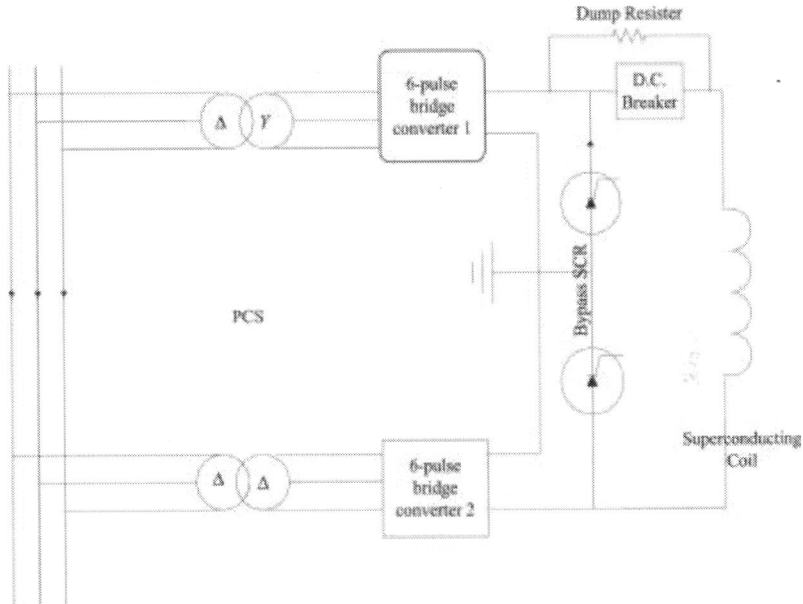

Figure 2. Schematic diagram of SMES connected to electric AC grid.

The control of the operation of SMES during its charging, discharging, and steady state mode and the power modulating dynamic oscillatory period are achieved by the application of adequate positive or negative voltage to the inductor, through the control of firing angle of the converter bridges. Fig. 3 illustrates the transfer function model representation of the SMES control scheme, where the ACE may be given to the proportional block (K_{SMES}) to derive the incremental change in converter voltage (ΔEd), as explained in Eq. (3). In order to achieve quick restoration of inductor current (Id) after any possible change in load demand in the system, the incremental ΔId is sensed and used as a negative feedback signal in the SMES control loop (Banerjee et al., 1990):

$$\Delta E_{d_i} = \frac{1}{1 + sT_{dc_i}}[K_{SMES}(\beta_i \Delta f_i + \Delta P_{ij}) - K_{Id_i} \Delta I_{d_i}]$$

$$(3)$$

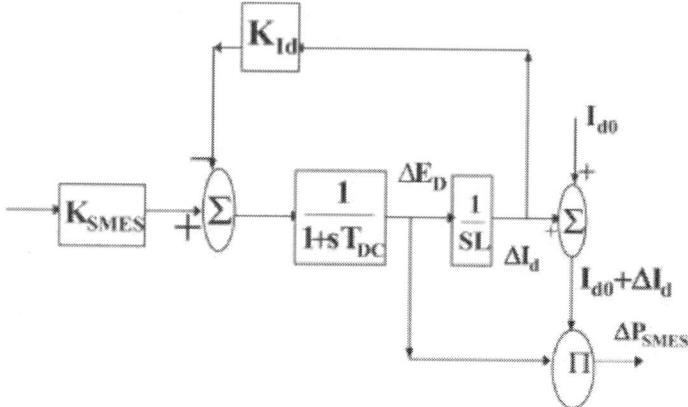

Figure 3. Transfer function model of SMES units.

In Eq. (3), Tdc is the converter time delay in s; K_{SMES} is the gain of the SMES control loop for ACE signal in kV/unit ACE, KId is the gain of the inductor current deviation feedback loop in kV/kA.

Since the amount of stored energy is finite, the inductor current falls. The deviation in the inductor current ΔId is expressed as follows in Eq. (4):

$$\Delta I_d = \frac{\Delta E_d}{PL}$$

(4)

where P is the differential operator with respect to time. The deviation in the inductor power flow ΔP_{SMES} is given by the expression as follows in Eq. (5):

$$\Delta P_{SMES} = I_{d0} \cdot \Delta E_d + \Delta E_d \cdot \Delta I_d$$

(5)

The inductor is initially charged to its rated current Id_0 by applying a small positive voltage. Once the current has attained the rated value, it is held constant by reducing the voltage ideally to zero since the coil is superconducting. However, a very small voltage may be required to overcome the commutating resistance.

THE FORMULATION OF THE PROBLEM AND OBJECTIVE FUNCTIONS

As the problem is planned to be formulated in an optimization framework, suitable design of objective function is tantamount to the efficacy of overall control performance. Hence, different objective functions, i.e., J_1 and J_2, integral of time multiple of absolute error (ITAE) and integral of time multiple of square of errors (ITSE) are discussed:

$$J_1 = \text{ITAE} = \int_0^{t_{sim}} t[|(\Delta f_1)| + |(\Delta f_2)| + |(\Delta P_{Tie})|] \cdot dt \tag{6}$$

In the above equation, *tsim* is the time range of simulation:

$$J_2 = \text{ITSE} = \int_0^{t_{sim}} t[(\Delta f_1)^2 + (\Delta f_2)^2 + (\Delta P_{Tie})^2] \cdot dt \tag{7}$$

The values of ITSE, settling time (*Ts*) of both area frequency deviations (Δf_1 and Δf_2) and $\Delta PTie$ along with the minimum damping ratios among all the system eigenvalues are combined to formulate the third objective function J_3 as given below in Eq. (8):

$$J_3 = \omega_1(\text{ITSE}) + \omega_2(1/X) + \omega_3(Ts) \tag{8}$$

ω_1, ω_2 and ω_3 are the weighing factors suitably chosen. x = minimum damping ratio (MDR) among all the eigenvalues of the system. Ts = settling time; $T_s = T_{s_{f_1}} + T_{s_{f_2}} + T_{s_{\Delta P_{Tie}}}$. $T_{s_{\Delta P_{Tie}}}$, T_{sf1}, T_{sf2}. Settling time of tie line power deviation, frequency deviation in area 1 and 2.

CUCKOO SEARCH ALGORITHM: AN OVERVIEW

CSA is an evolutionary algorithm that is inspired by the *brood parasitism* found in the breeding behaviour of some commonly found species of Cuckoos. Moreover, the algorithm also incorporates into its structure the mathematical model of the behaviour of Lévy flight found in some birds and fruit flies (Yang and Deb, 2009).

In the evolution strategy of CSA, an important methodology of choosing a random direction to generate the *step length* is done with the help of an algorithm known as *Mantegna* algorithm. The above process guarantees a stable symmetric Lévy distribution (Yang, 2010).

Cuckoo search algorithm: the programming methodology

As far as applying the algorithm is concerned, three simplifying assumptions described below have been used in this work.

(i) Each cuckoo lays one egg at a time which it dumps in a randomly selected nest (n).

(ii) The best nests having better quality eggs are retained for subsequent generations.

(iii) Keeping the total numbers of host nests as constant, an egg laid by a cuckoo could be detected by the host bird with a probability (Pa) of 0.1.

Based on these three rules, the basic steps of the CSA are provided in the flow chart shown in Fig. 4.

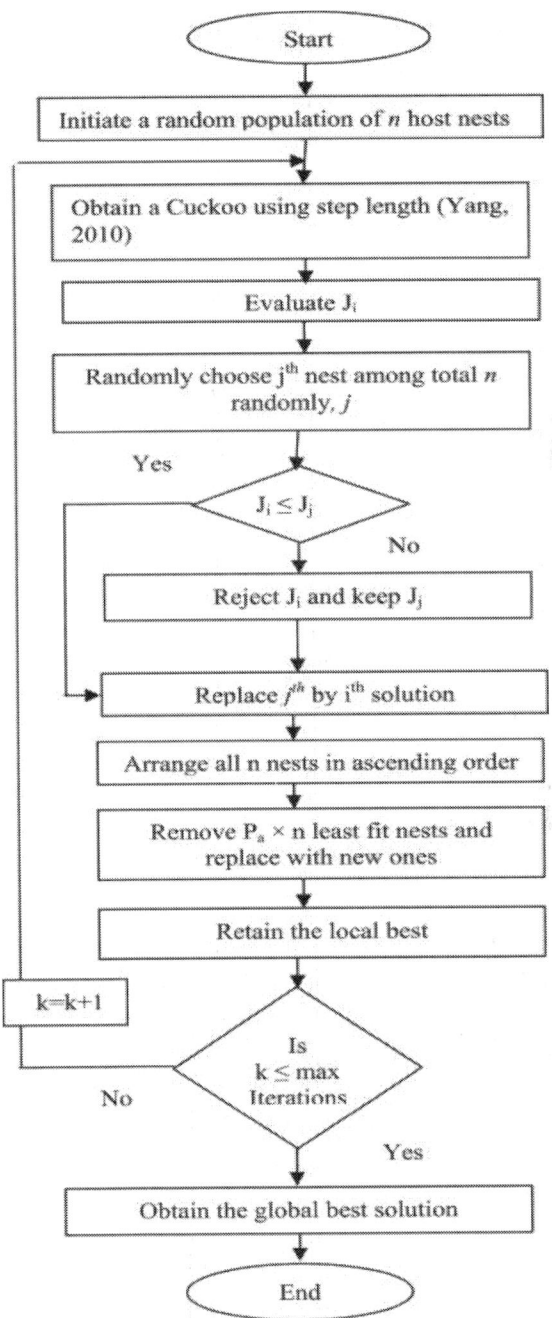

Figure 4. Flowchart of conventional Cuckoo search algorithm.

SIMULATION AND RESULTS

The AGC system model developed in MATLAB/SIMULINK is used to obtain dynamic response for a step load perturbation. The integral controller parameter (KI) for the main AGC loop, SMES gain parameter (K_{SMES}) and feedback gain (KId) in SMES control loop are to be obtained separately by optimizing the three different objective functions. For the purpose of optimization Cuckoo search algorithm is applied.

Two numbers of SMES each having capacities of 30 MJ are incorporated in both the areas. The actual and per unit value of SMES device are given in Appendix A.

Moreover, in order to take into account the smallest time constants associated with SMES, time-domain analysis of the continuous system is performed with a time step of 0.01 s with appropriate choice of sampling time intervals for the controllers.

CSA tuned controller parameters (KI, K_{SMES}, KId)

For optimizing the objective functions each of the parameters are randomly initialized in suitable ranges and the parameters evolve through successive generation giving the optimum results at the end. The values of all the controller parameters are obtained separately by optimizing the objective functions J_1, J_2 and J_3 with the help of CSA in three different runs of the algorithm. It is to be noted that the objective functions are dependent on the time domain and eigenvalue based performance indices (PFIs), which are evaluated at the end of simulation time of 30 s. All these PFIs are also evaluated for the case when none of the either areas is operating with SMES. The optimized values of the above mentioned controller parameters obtained separately by optimizing the three objective functions are elucidated in Table 1.

Table 1. The optimized values of controller parameters with the three objective functions.

Objective function/controller parameters	Controller parameters			Optimized value of objective function
	SMES loop gain (K_{SMES})	Inductor current feedback gain (KId)	Integral gain (KI)	
Different objective functions tuned in CSA with SMES				
J_1	85.8402	20.1975	5.0464	0.0423
J_2	91.9804	1.7957	3.4621	2.5732e−004
J_3	99.1703	14.3716	4.7118	14.843
J_3 tuned in PSO with SMES	96.097	19.2532	4.9322	17.0554
J_3 tuned in CSA without SMES	0	0	0.4986	76.3698

Controller performance evaluation from optimization results

Besides the PFIs defined and used in the formulation of the objective functions, two other PFIs, i.e., integral square error (ISE) and integral absolute error (IAE) are also evaluated and compared for each set of optimized controllers. These PFIs are enumerated in Table 2. From the results, it is clear that with the proposed algorithm the system modes shift more in the left half of S-plane, which enhances the system stability. Minimum damping ratio (MDR) of system among all the system eigenvalues, obtained separately for all the objective functions optimized with CS are also illustrated. The settling of deviations is very fast, around 5 s in objective function J_3 and damping ratio also improved as compared to other objective functions in Table 2.

Table 2. Several PFIs and MDR among all the eigenvalues of the system using CSA based on different objective functions J_1, J_2, and J_3 with and without SMES.

Performance indices	Different objective functions with SMES			J_3 without SMES
	J_1	J_2	J_3	
ISE	3.5351e−004	2.8043e−004	3.1536e−004	0.0017
ITSE	4.2914e−004	2.5732e−004	3.5042e−004	0.0031
IAE	0.0312	0.0365	0.0324	0.1252
ITAE	0.0423	0.0758	0.0484	0.451
Ts (s)				
$\Delta f1$	10.35	10.35	4.9	25.49
$\Delta f2$	11.54	11.54	4.34	26.58
$\Delta PTie$	9.34	9.34	3.98	23.98
MDR	0.6227	0.6561	0.6954	0.0098
Eigenvalues	−12.5000−26.0324	−12.5000−26.4451	−12.5000−24.2947	−3.3333−12.5000
	−3.1790 + 3.9950i	−3.4691 + 3.9902i	−3.8294 + 3.9570i	−0.0250 + 2.5572i
	−3.1790 − 3.9950i	−3.4691 − 3.9902i	−3.8294 − 3.9570i	−0.0250 − 2.5572i

Comparison of CSA with particle swarm optimization (PSO)

A comparison is also sought in this work between CSA and the widely accepted optimization technique PSO in terms of the performance obtained by the respective controllers when the gains of the same are obtained by optimizing the objective function J_3. As depicted in Fig. 5, Fig. 6 and Fig. 7, which show both the areas' frequency and tie line power deviations for 0.01 SLP in the 1st area, CSA tuned controller has provided better damping compared to the one tuned by PSO. Moreover, looking at the several performance indices given in Table 3, it can be seen that with CSA optimized controller, the performances particularly the settling times are better compared to those obtained with PSO.

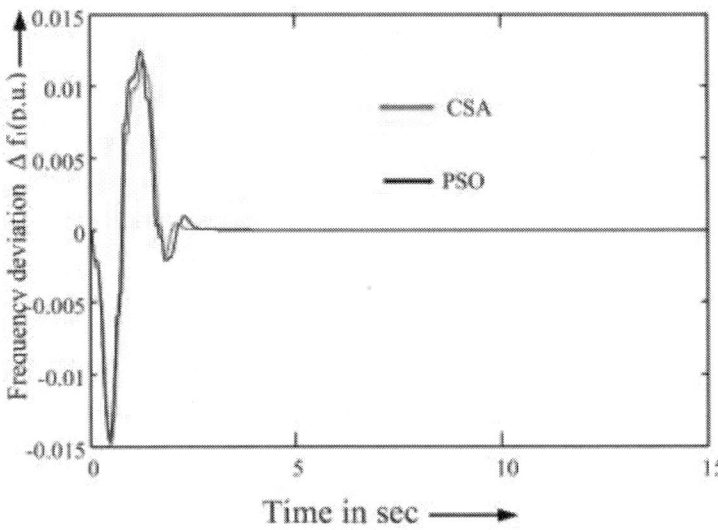

Figure 5. Change in frequency of 1st area for 1% load change in 1st area.

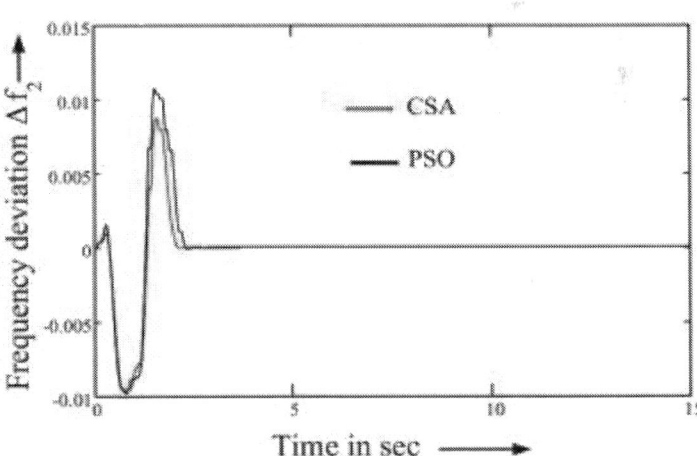

Figure 6. Change in frequency of 2nd area for 1% load change in 1st area.

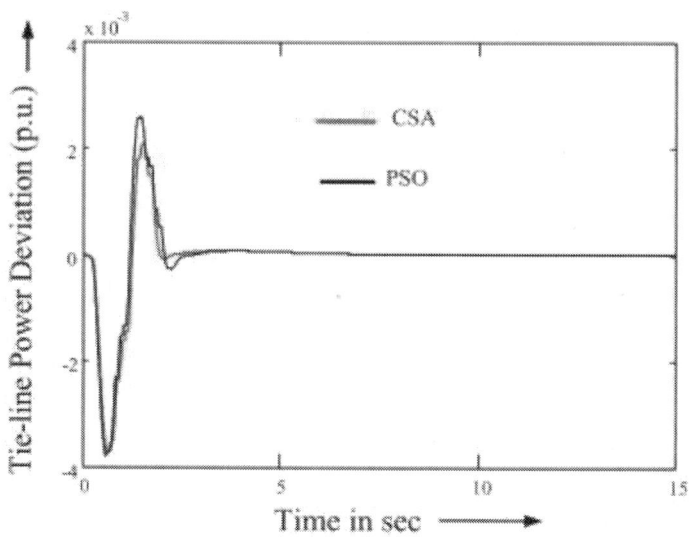

Figure 7. Change in Tie-line power for 1% load change in 1st area.

Table 3. Comparison of several PFIs and MDR of the system with controllers tuned with CSA and PSO based on objective function J_3.

Performance indices	ISE	ITSE	Ts			MDR
			Δf_1	Δf_2	ΔPTi_e	
CSA tuned J_3	3.1536e−004	3.5042e−004	4.9	4.3	3.98	0.695
PSO tuned J_3	3.3400e−004	3.8993e−004	5.7	5.1	5.18	0.679

Controller performance evaluation for different disturbances and changes in parameters

Step load increase in area 1

The Integral and SMES controller parameters are set at the values obtained by optimizing objective function J_3. The optimal value for the integral gain KI found in both cases without and with SMES is given in Table 1. The first control area is subjected to a SLP of 1% from its nominal value at time $t = 0$. Dynamic response of frequency deviations (Δf_1 and Δf_2) of both the control areas and the deviation in tie line power ($\Delta P Tie$) obtained for this perturbation is depicted in Fig. 8, Fig. 9 and Fig. 10. From the figures it can be witnessed that, the frequency and tie line power oscillations are seen to settle around 25 s without SMES, whereas the same value reduces to 5 s with the SMES operating. The values of over-shoot, under-shoot and MDR have also improved predominantly as enumerated in Table 2.

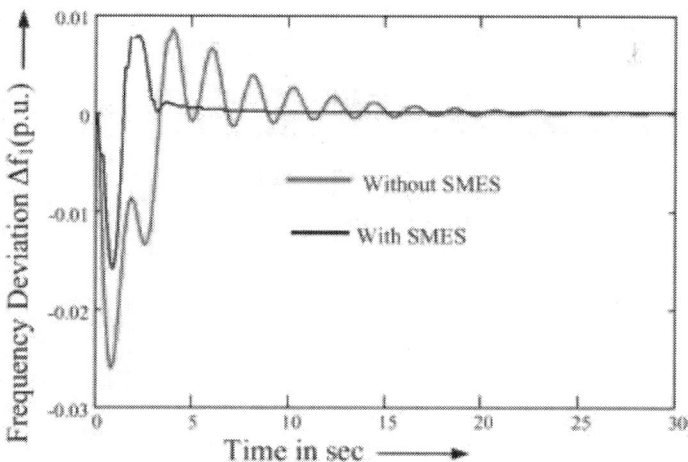

Figure 8. Change in frequency of 1st area for 1% load change in 1st area.

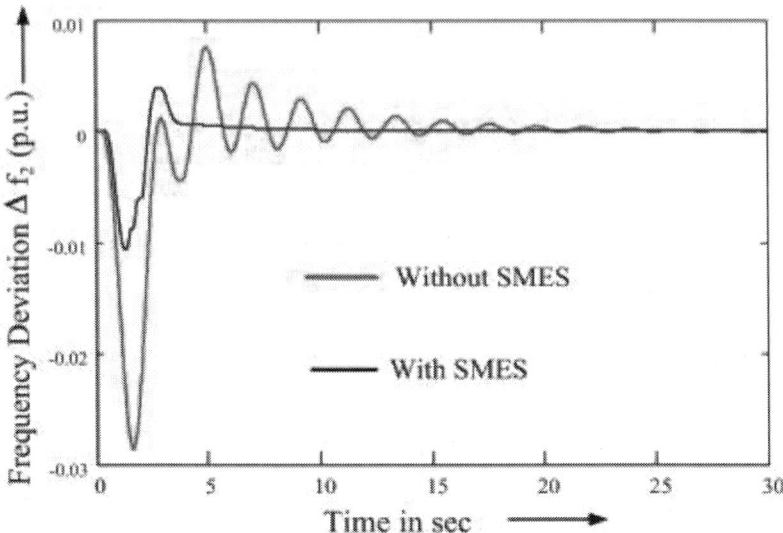

Figure 9. Change in frequency of 2nd area for 1% load change in 1st area.

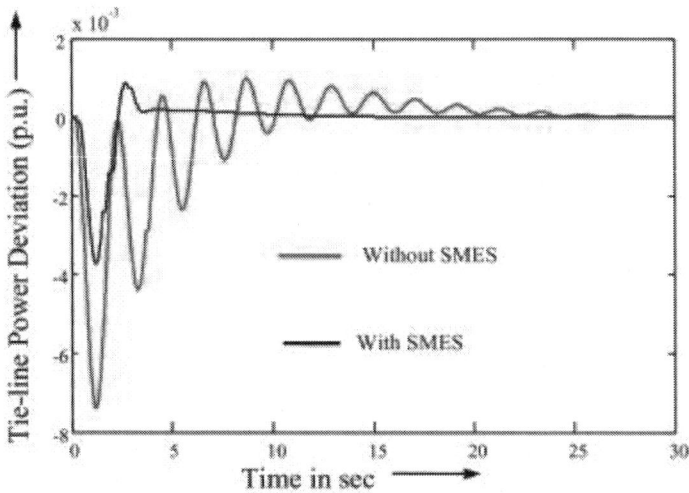

Figure 10. Change in tie-line power for 1% load change in 1st areas.

Step load increase in both area 1 and 2
Both the control areas are subjected to SLP of 1% each from their nominal values as it was done in the previous two cases. Dynamic

response of frequency deviations (Δf_1 and Δf_2) and the deviation in tie line power ($\Delta PTie$) obtained for this perturbations are depicted in Fig. 11, Fig. 12 and Fig. 13. From Fig. 11 and Fig. 12, it can be seen that the both the area frequency deviations become more when loads in both of them are increased simultaneously. Moreover, it can be noticed from Fig. 13 that, when both the areas are subjected to same step load perturbation, frequency deviations in both the areas have increased, but no oscillation is noticed in tie line power deviation as equal disturbances given to two equal area having same inertia would not require any exchange through tie line.

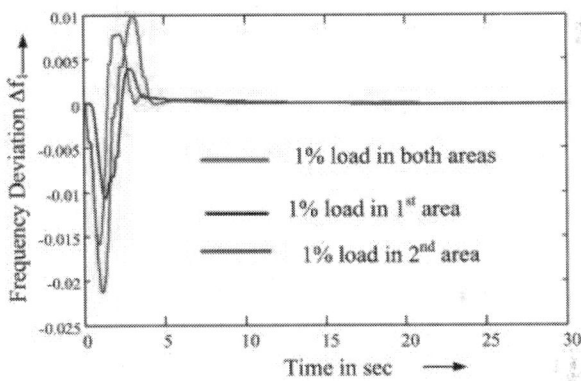

Figure 11. Change in frequency of 1st area due to load variation in 1st, 2nd and both areas.

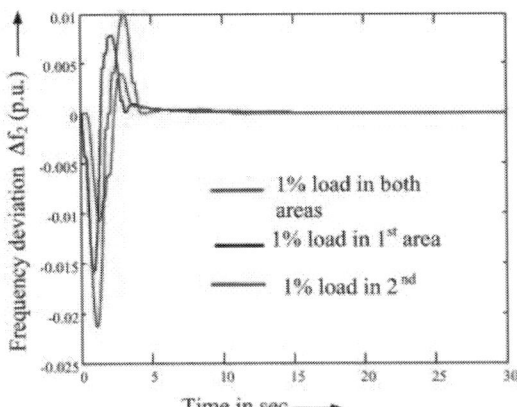

Figure 12. Change in frequency of 2nd area due to load variation in 1st, 2nd and both areas.

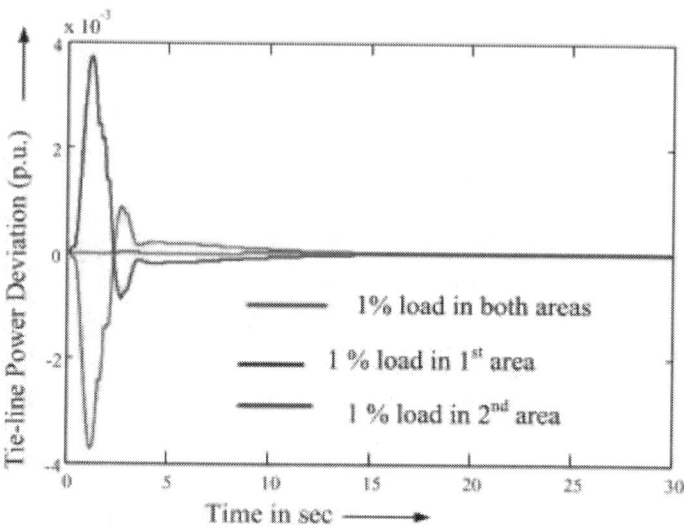

Figure 13. Change in tie-line power due to load variation in 1st, 2nd and both areas.

Sensitivity analysis with variation in parameter

To study the robustness of the proposed controllers obtained by optimizing J_3 variations in the system parameters and operating conditions are deliberately introduced. For testing the controller performance with parameter variations, several numbers of time constants related to the governing system (Tg), turbines (Tt), the synchronizing power coefficient (T_{12}), reheater (Tr) are varied in the range of -50% to $+50\%$ from their respective nominal values in separate events of perturbation cases. Similar variations in the values of reheater gain (Kr) and inertia constant (H) have also not disturbed the oscillations keeping them stable.

Moreover, the numerical values of ITSE, frequency deviations of both the areas and the tie line power deviations obtained with variations in system parameters and operating conditions are listed in Table 4. The values obtained corroborate the robustness of the proposed controller for modified parameters and operating conditions.

Table 4. Sensitivity analysis.

Parameter variation	% change	Performance index ITSE	Settling time			MDR
			Δf_1	Δf_2	ΔP_{Tie}	
T_{12}	+50	2.0071e−004	10.6400	11.4500	9.6500	0.5020
	+25	2.2153e−004	10.6100	11.5900	9.6400	0.5681
	−25	3.1676e−004	10.4900	11.9900	9.3700	0.6828
	−50	4.0349e−004	10.5300	12.5100	8.9500	0.7364
T_g	+50	3.2611e−004	11.9300	11.9300	9.4800	0.6954
	+25	2.8919e−004	10.7500	11.9300	9.6200	0.6954
	−25	2.3025e−004	10.4500	11.6400	9.5000	0.6954
	−50	2.0743e−004	10.8900	11.8500	9.8400	0.6954
T_r	+50	4.0266e−004	13.3000	14.5500	14.1400	0.6954
	+25	3.2122e−004	12.1700	13.4500	12.1900	0.6954
	−25	2.0633e−004	7.4000	8.3100	3.5500	0.6954
	−50	1.5441e−004	4.6900	4.7700	3.7900	0.6954
K_r	+50	1.8588e−004	3.8300	4.3100	3.4500	0.6954
	+25	2.1237e−004	8.5600	9.5300	4.5300	0.6954
	−25	3.5564e−004	10.8200	11.9800	11.9000	0.6954
	−50	5.8164e−004	19.5000	20.2500	12.2500	0.6954
T_t	+50	4.1800e−004	10.7400	11.8300	9.0600	0.6954
	+25	3.3124e−004	10.6500	11.8100	9.4500	0.6954
	−25	1.9742e−004	10.5600	11.7400	9.5400	0.6954
	−50	1.4947e−004	10.8600	12.0100	9.6200	0.6954
Loading condition	+50	1.5537e−004	10.8000	11.9800	9.6900	0.6006
	+25	1.5261e−004	10.8400	12.0100	9.6700	0.6194
	−25	1.4640e−004	12.4900	12.4900	10.0200	0.7234
	−50	1.4685e−004	11.4100	12.5300	10.0300	0.8035
H	+50	2.7482e−004	10.4300	11.6100	9.2800	0.5994
	+25	2.6620e−004	10.4400	11.6300	9.3700	0.6187
	−25	2.5164e−004	10.7800	11.9700	9.8500	0.7240
	−50	2.5466e−004	10.8600	12.0100	10.0200	0.8041

CONCLUSION

In this work, observed that effectively tuned controller gains of SMES along with those of the power system enable the later to operate in a more stable manner compared to the case when no SMES were present. It was found that, besides the efficiency of CSA optimization algorithm, suitable design of the objective function also plays an important role in obtaining a robust design of different controllers in a coordinated manner. However, any possible modification of the algorithm either through the process of hybridization with other similar evolutionary algorithms, or by altering the basic process for multi objective optimization problems may result in improving its efficiency.

APPENDIX A.

A.1. SMIB data

Total rated area capacity (Pr) = 2000 MW, f = 60 Hz, $R_1 = R_2 =$ 2.4 (Hz/p.u. MW), $Tt_1 = Tt_2 = 0.3$ s.

$Tr_1 = Tr_2 = 10$ s, $Tg_1 = Tg_2 = 0.08$ s, $Tp_1 = Tp_2 = 20$ s $(Tp = (2H/fD))$, $Kp_1 = Kp_2 = 120$ Hz/p.u. MW $(Kp = (1/D))$

$D_i = \Delta P_{D_i}/\Delta f_i = 8.33 \times 10^{-3}$ p.u. MW/Hz, $Kr_1 = Kr_2 = 0.5$, $\beta_1 = \beta_2 = 0.425$, $T_{12} = 0.0867$ s.

A.2. SMES system data

$Tdc_1 = Tdc_2 = 0.03$ s, SB = base power = 2000 MW, assuming base value of Ed = 10 kV and Id = 200 kA.

Base impedance $(ZBase)$ = 0.05 Ω, $L_1 = L_2 = 2.65$ H (absolute value) = 19,970 p.u.

The initial current Id_0 = 4.5 kA = 0.02 p.u. (see current base).

A.3. PSO parameters

Number of particles = 20, C_1 = 1.2, C_2 = 1.2, moment of inertia = 0.9.

Maximum number of step = 20, dimension of the problem = 3.

REFERENCES

1. Banerjee, S., Chatterjee, J.K., Tripathy, S.C., 1990. Application of magnetic energy storage unit as load frequency stabilizer. IEEE Trans. EnergyConvers. 5 (1), 46–51.
2. Bhatt, P., Ghoshal, S.P., Roy, R., 2010. Load frequency stabilization by coordinated control of thyristor controlled phase shifters and superconducting magnetic energy storage for three types of interconnected two-area power systems. Electr. Power Energy Syst. 32 (10), 1111–1124.

3. Elgerd, O.I., 2005. Electric Energy Systems Theory: An Introduction, 2nd ed., 25th reprint. McGraw-Hill.

4. Ghoshal, S.P., 2004. Optimizations of PID gains by particle swarm optimizations in fuzzy based automatic generation control. Electr. Power Syst. Res. 72, 203–212.

5. Ibraheem, P.K., Kothari, D.P., 2005. Recent philosophies of automatic generation control strategies in power systems. IEEE Trans. Power Syst. 20 (1), 346–357.

6. Mallesham, G., Mishra, S., Jha, A.N., 2010. Optimization of control parameters in AGC of microgrid using gradient descent method. In: 16th National Power Systems Conference, pp. 37–42.

7. Nanda,J., Mangla, A., Suri, S., 2006. Some new findings on automatic generation control of an interconnected hydrothermalsystem with conventional controllers. IEEE Trans. Energy Convers. 21 (1), 87–194.

8. Tripathy, S.C., Balasubramania, R., Chanramohanan, N.P.S., 1992. Adaptive automatic generation control with superconducting magnetic energy storage in power system. IEEE Trans. Energy Convers. 7 (3), 434–441.

9. Yang, X.S., Deb, S.,2009. Cuckoo search via Lévy flights. In: Proc. of World Congress on Nature & Biologically Inspired Computing (NaBIC 2009), India. IEEE Publications, USA.

10. Yang, X.S., 2010. Nature-Inspired Metaheuristic Algorithms, 2nd ed. Luniver Press

CITATION

Sabita Chaine, M. Tripathy, Design of an optimal SMES for automatic generation control of two-area thermal power system using Cuckoo search algorithm, Journal of Electrical Systems and Information Technology, Volume 2, Issue 1, May 2015, Pages 1-13, ISSN 2314-7172, http:// dx.doi.org/10.1016/j.jesit.2015.03.001.

CHAPTER 5

Design and Analysis of Differential Evolution Algorithm Based Automatic Generation Control For Interconnected Power System

Umesh Kumar Rout, Rabindra Kumar Sahu, Sidhartha Panda,

Technology (VSSUT), Burla 768 018, Odisha, India

ABSTRACT

This paper presents the design and performance analysis of Differential Evolution (DE) algorithm based Proportional-Integral (PI) controller for Automatic Generation Control (AGC) of an interconnected power system. A two area non-reheat thermal system equipped with PI controllers which is widely used in literature is considered for the design and analysis purpose. The design problem is formulated as an optimization problem control and DE is employed to search for optimal controller parameters. Three different objective functions using Integral Time multiply Absolute Error (ITAE), damping ratio of dominant eigenvalues and settling time with appropriate weight coefficients are derived in order to increase the performance of the controller. The superiority of the proposed DE optimized PI controller has been shown by comparing the results with some recently published modern heuristic optimization techniques such as Bacteria Foraging

Optimization Algorithm (BFOA) and Genetic Algorithm (GA) based PI controller for the same interconnected power system.

INTRODUCTION

An interconnected power system is made up of several areas and for the stable operation of power systems; both constant frequency and constant tie-line power exchange should be provided. In each area, an Automatic Generation Controller (AGC) monitors the system frequency and tie-line flows, computes the net change in the generation required (generally referred to as Area Control Error – ACE) and changes the set position of the generators within the area so as to keep the time average of the ACE at a low value [1]. Therefore ACE, which is defined as a linear combination of power net-interchange and frequency deviations, is generally taken as the controlled output of AGC. As the ACE is driven to zero by the AGC, both frequency and tie-line power errors will be forced to zeros[2]. AGC function can be viewed as a supervisory control function which attempts to match the generation trend within an area to the trend of the randomly changing load of the area, so as to keep the system frequency and the tie-line power flow close to scheduled value. The growth in size and complexity of electric power systems along with increase in power demand has necessitated the use of intelligent systems that combine knowledge, techniques and methodologies from various sources for the real-time control of power systems.

The researchers in the world over trying to understand several strategies for AGC of power systems in order to maintain the system frequency and tie line flow at their scheduled values during normal operation and also during small perturbations. A critical literature review on the AGC of power systems has been presented in [3] where various control aspects concerning AGC problem have been studied. Moreover the authors have reported various AGC schemes, AGC strategies and AGC system incorporating BES/SMES, wind turbines, FACTS devices and PV systems. There has been considerable research work attempting to propose better AGC systems based on modern control theory [4] and [5], artificial neural network [6], [7], [8] and [9], fuzzy system theory [10], [11] and [12], reinforcement learning [13] and ANFIS approach has[14] and [15].

From the literature survey, it may be concluded that there is still scope of work on the optimization of PI controller parameters to further improve the AGC performance. For this, various novel evolutionary optimization techniques can be proposed and tested for comparative optimization performance study. However, ANN, fuzzy, and ES suffer from the requirement of expert user in their design and implementation, a lack of the formal model theory and mathematical rigors and so are vulnerable to the experts' depth of knowledge in problem definition. Modern heuristic optimization techniques, by contrast, access deep knowledge of systems problem by well-established models and have much more potential in power systems. Modern heuristic optimization technique based approaches have been proposed recently to design a controller. These approaches include particle swarm optimization [16] and [17], differential evolution [18] and [19], multi-objective evolutionary algorithm [20] and NSGA-II [21] and [22], etc. Nanda et al.[23] have demonstrated that bacterial foraging, a powerful evolutionary computational technique, based integral controller provides better performance as compared to that with integral controller based on classical and GA techniques in three unequal areas thermal system. E.S. Ali and S.M. Abd-Elazim [24] have reported recently that Bacterial Foraging Optimization Algorithm (BFOA), based proportional integral (PI) controller provides better performance as compared to that with GA based PI controller in two area non-reheat thermal system. Differential Evolution (DE) is a branch of evolutionary algorithms developed by Rainer Stron and Kenneth Price in 1995 for optimization problems [25]. It is a population-based direct search algorithm for global optimization capable of handling non-differentiable, non-linear and multi-modal objective functions, with few, easily chosen, control parameters. It has demonstrated its usefulness and robustness in a variety of applications such as, Neural network learning, Filter design and the optimization of aerodynamics shapes. DE differs from other evolutionary algorithms (EA) in the mutation and recombination phases. DE uses weighted differences between solution vectors to change the population whereas in other stochastic techniques such as Genetic Algorithm (GA) and expert systems (ES), perturbation occurs in accordance with a random quantity. DE employs a greedy selection process with inherent elitist features. Also it has a minimum number of control parameters, which can be tuned effectively[18] and [19]. In view of the above, an attempt has been made in this paper for the optimal design of DE based PI controller for LFC in two area interconnected power system considering small step load perturbation occurring in a single area as well as simultaneously in all the areas.

The aim of the present work is twofold: to demonstrate the advantages of DE over other techniques such as BFOA and GA which are recently presented in the literature for the similar problem and to show advantages of using a modified objective function based on Integral Time multiply Absolute Error (ITAE) criteria, damping ratio of dominant eigenvalues and settling times of frequency and tie line power deviations with appropriate weight coefficients to further increase the performance of the proposed controllers. The design problem of the proposed controller is formulated as an optimization problem and DE is employed to search for optimal controller parameters. By minimizing the proposed objective functions, in which the deviations in the frequency and tie line power, damping ratio and settling times are involved; dynamic performance of the system is improved. Simulations results are presented to show the effectiveness of the proposed controller in providing good damping characteristic to system oscillations over a wide range of loading conditions, disturbance and system parameters. Further, the superiority of the proposed design approach is illustrated by comparing the proposed approach with some recently published approaches such as BFOA and GA.

SYSTEM MODELING

LFC model

The dynamic model of Load Frequency Control (LFC) for a two-area interconnected power system is presented in this section. Each area of the power system consists of speed governing system, turbine and generator. Each area has three inputs and two outputs. The inputs are the controller input ΔP_{ref} (also denoted as u), load disturbance ΔP_D and tie-line power error ΔP_{Tie}. The outputs are the generator frequency Δf and Area Control Error (ACE) given by Eq. (1).

$$ACE = B\Delta f + \Delta PTie \tag{1}$$

where B is the frequency bias parameter.

To simplicity the frequency-domain analyses, transfer functions are used to model each component of the area. Turbine is represented by the transfer function [2]:

$$G_T(s) = \frac{\Delta P_T(s)}{\Delta P_V(s)} = \frac{1}{1 + sT_T} \tag{2}$$

From [2], the transfer function of a governor is:

$$G_G(s) = \frac{\Delta P_V(s)}{\Delta P_G(s)} = \frac{1}{1 + sT_G} \tag{3}$$

The speed governing system has two inputs ΔP_{ref} and Δf with one out put $\Delta P_G(s)$ given by [2]:

$$\Delta P_G(s) = \Delta P_{ref}(s) - \frac{1}{R}\Delta f(s) \tag{4}$$

The generator and load is represented by the transfer function [2]:

$$G_P(s) = \frac{K_P}{1 + sT_P} \tag{5}$$

where $K_P = 1/D$ and $T_P = 2H/fD$.

The generator load system has two inputs $\Delta P_T(s)$ and $\Delta P_D(s)$ with one out put $\Delta f(s)$ given by [2]:

$$\Delta f(s) = {}_{GP}(s)[\Delta_{PT}(s) - \Delta_{PD}(s)] \qquad (6)$$

System under study

The system under investigation consists of two area interconnected power system of nonreheat thermal plant as shown in Fig. 1. The system is widely used in literature is for the design and analysis of automatic load frequency control of interconnected areas [24]. In Fig. 2, B_1 and B_2 are the frequency bias parameters; ACE_1 and ACE_2 are area control errors; u_1 and u_2 are the control outputs form the controller; R_1 and R_2 are the governor speed regulation parameters in p.u. Hz; T_{G1} and T_{G2} are the speed governor time constants in s; ΔP_{V1} and ΔP_{V2} are the change in governor valve positions (p.u.); ΔP_{G1} and ΔP_{G2} are the governor output command (p.u.); T_{T1} and T_{T2} are the turbine time constant in s; ΔP_{T1} and ΔP_{T2} are the change in turbine output powers; ΔP_{D1} and ΔP_{D2} are the load demand changes; ΔP_{Tie} is the incremental change in tie line power (p.u.); K_{PS1} and K_{PS2} are the power system gains; T_{PS1} and T_{PS2} are the power system time constant in s; T_{12} is the synchronizing coefficient and Δf_1 and Δf_2 are the system frequency deviations in Hz. The relevant parameters are given in Appendix A.

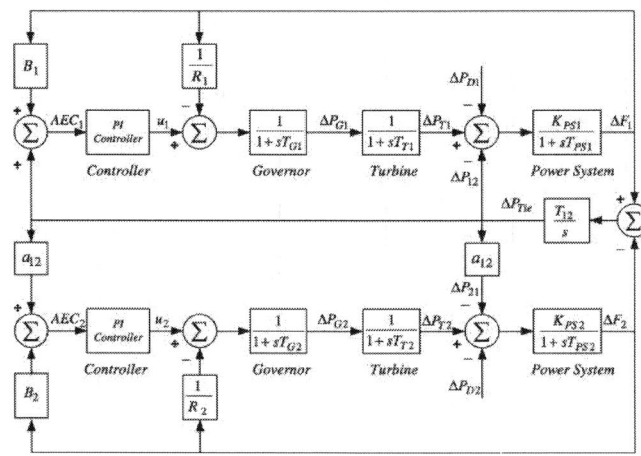

Figure 1. Transfer function model of two-area nonreheat thermal system.

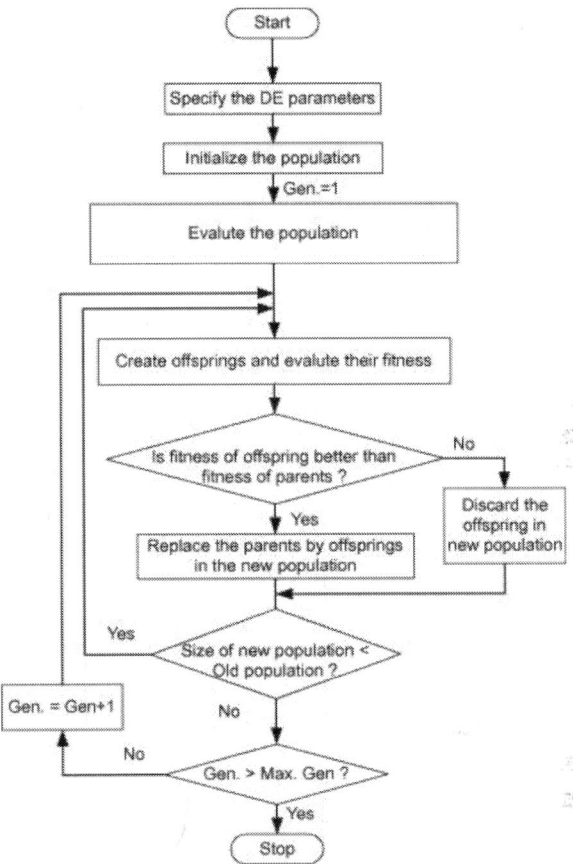

Figure 2. Flow chart of proposed DE optimization approach.

THE PROPOSED APPROACH

The proportional integral derivative controller (PID) is the most popular feedback controller used in the process industries. While proportional and integrative modes are often used as single control modes, a derivative mode is rarely used as it amplifies the signal noise. In view of the above, a PI structured controller is considering in the present paper. The design of PI controller requires determination of the two parameters, Proportional gain (K_P) and Integral gain (K_I). The controllers in both the areas are considered to be identical so that $K_{P1} = K_{P2} = K_P$ and $K_{I1} = K_{I2} = K_I$.

The error inputs to the controllers are the respective Area Control Errors (ACEs) given by:

$$e_1(t) = ACE_1 = B_1\Delta f_1 + \Delta PTie \tag{7}$$

$$e_2(t) = ACE_2 = B_2\Delta f_2 - \Delta PTie \tag{8}$$

The control inputs of the power system u_1 and u_2 are the outputs of the controllers. The control inputs are obtained as:

$$u_1 = K_{P1}ACE_1 + K_{I1}\int ACE_1 \tag{9}$$

$$u_2 = K_{P2}ACE_2 + K_{21}\int ACE_2 \tag{10}$$

In the design of a PI controller, the objective function is first defined based on the desired specifications and constraints. The design of objective function to tune PI controller is generally based on a performance index that considers the entire closed loop response. Typical output specifications in the time domain are peak overshooting, rise time, settling time, and steady-state error. Four kinds of performance criteria usually considered in the control design are the Integral of Time multiplied Absolute Error (ITAE), Integral of Squared Error (ISE), Integral of Time multiplied Squared Error (ITSE) and Integral of Absolute Error (IAE). It has been shown that the ITAE provides better responses as compared to other criteria [26]. Also, the eigenvalues and modal analysis provides an extension of analytical methods to examine the low frequency oscillations that are present in a power system. Eigenvalue analysis uses the standard linear, state space form of system equations and provides an appropriate tool for evaluating system conditions for the study of small signal stability of power system. Eigenvalue analysis investigates the dynamic behavior of the power system under different characteristic frequencies (modes). In a power system, it is required that all modes be stable. Moreover, it is desired that all electromechanical oscillations be damped out as quickly as possible. In other words, the damping ratios of dominant eigenvalues should be maximized as much as possible [1].

Some of the realistic control specifications for Automatic Generation Control (AGC) are[2]:

(i) The frequency error should return to zero following a load change.
(ii) The integral of frequency error should be minimum.
(iii) The control loop must be characterized by a sufficient degree of stability.
(iv) Under normal operating conditions, each area should carry its own load.

To meet the above design specifications, three different objective functions are employed in the present paper as given by Eqs. (11), (12) and (13). The first objective function (ITAE) given by Eq. (11) is a standard one and employed often in other papers. It tries to achieve the design specifications given by (i and ii). Also, the desired system response should have minimal settling time with a small or no overshoot. To add some degree of stability and damping of oscillating modes, the second objective function given by Eq. (12) is proposed. It aims to minimize the ITAE and maximize minimum damping ratio of dominant eigenvalues. Minimization of this objective function will minimize maximum overshoot also [10] and [11]. To ensure that the errors are quickly minimized the settling times of Δf_1, Δf_2 and ΔP_{Tie} are also included in the third objective function given by Eq. (13).

$$J_1 = \int_0^{t_{sim}} (|\Delta f_1| + |\Delta f_2| + |\Delta P_{Tie}|) \cdot t \cdot dt \tag{11}$$

$$J_2 = \int_0^{t_{sim}} \omega_1 \cdot \int_0^{t_{sim}} (|\Delta f_1| + |\Delta f_2| + |\Delta P_{Tie}|) \cdot t \cdot dt + \omega_2$$
$$\cdot \frac{1}{\min \left(\sum_{i=1}^n (1 - \zeta_i) \right)} \tag{12}$$

$$J_3 = \int_0^{t_{sim}} \omega_3 \cdot \int_0^{t_{sim}} (|\Delta f_1| + |\Delta f_2| + |\Delta P_{Tie}|) \cdot t \cdot dt + \omega_4$$

$$\cdot \frac{1}{\min \left(\sum_{i=1}^{n}(1 - \zeta_i) \right)} + \omega_5 \cdot T_S \tag{13}$$

where Δf_1 and Δf_2 are the system frequency deviations; ΔP_{Tie} is the incremental change in tie line power; t_{sim} is the time range of simulation; ζ_i is the damping ratio and n is the total number of the dominant eigenvalues; T_S is the sum of the settling times of frequency and tie line power deviations; ω_1 to ω_5 are weighting factors. Inclusion of appropriate weighting factors to the right hand individual terms helps to make each term competitive during the optimization process. Wrong choice of the weighting factors leads to incompatible numerical values of each term involved in the definition of fitness function which gives misleading result. The weights are so chosen that numerical value of all the terms in the right hand side of Eqs. (12) and (13) lie in the same range. Repetitive trial run of the optimizing algorithms reveals that numerical value of ITAE lies in the range 1.25–2.5, minimum damping ratio lies in the range 0.02–0.3 and total settling times of Δf_1, Δf_2 and ΔP_{Tie} lies in the range 15–50. To make each term competitive during the optimization process the weights are chosen as: $\omega_1 = 1.0$, $\omega_2 = 10$, $\omega_3 = 1.0$, $\omega_4 = 1.0$ and $\omega_5 = 0.05$.

The problem constraints are the PI controller parameter bounds. Therefore, the design problem can be formulated as the following optimization problem.

$$\text{Minimize} J \tag{14}$$

Subject to

$$K_{Pmin} \leqslant K_P \leqslant K_{Pmax}, \quad K_{Imin} \leqslant K_I \leqslant K_{Imax} \tag{15}$$

where J is the objective function (J_1, J_2 and J_3) and K_{Pmin}, K_{Imin} and K_{Pmax}, K_{Imax} are the minimum and maximum value of the control parameters. As reported in the literature, the minimum and maximum values of controller parameters are chosen as −1.0 and 1.0 respectively.

DIFFERENTIAL EVOLUTION

Differential Evolution (DE) algorithm is a population-based stochastic optimization algorithm recently introduced [25]. Advantages of DE are: simplicity, efficiency and real coding, easy use, local searching property and speediness. DE works with two populations; old generation and new generation of the same population. The size of the population is adjusted by the parameter N_P. The population consists of real valued vectors with dimension D that equals the number of design parameters/control variables. The population is randomly initialized within the initial parameter bounds. The optimization process is conducted by means of three main operations: mutation, crossover and selection. In each generation, individuals of the current population become target vectors. For each target vector, the mutation operation produces a mutant vector, by adding the weighted difference between two randomly chosen vectors to a third vector. The crossover operation generates a new vector, called trial vector, by mixing the parameters of the mutant vector with those of the target vector. If the trial vector obtains a better fitness value than the target vector, then the trial vector replaces the target vector in the next generation. The evolutionary operators are described below[18] and [19].

Initialization

For each parameter j with lower bound X_j^L and upper bound X_j^U, initial parameter values are usually randomly selected uniformly in the interval $[X_j^L, X_j^U]$.

Mutation

For a given parameter vector $X_{i,G}$, three vectors ($X_{r1,G}$ $X_{r2,G}$ $X_{r3,G}$) are randomly selected such that the indices i, $r1$, $r2$ and $r3$ are distinct. A donor vector $V_{i,G+1}$ is created by adding the weighted difference between the two vectors to the third vector as:

$$V_{i,G+1} = X_{r1,G} + F \cdot (X_{r2,G} - X_{r3,G}) \tag{16}$$

where F is a constant from $(0, 2)$

Crossover

Three parents are selected for crossover and the child is a perturbation of one of them. The trial vector $U_{i,G+1}$ is developed from the elements of the target vector ($X_{i,G}$) and the elements of the donor vector ($X_{i,G}$). Elements of the donor vector enters the trial vector with probability CR as:

$$U_{j,i,G+1} = \begin{cases} V_{j,i,G+1} & \text{if} \quad rand_{j,i} \leqslant CR \quad \text{or} \quad j = I_{rand} \\ X_{j,i,G+1} & \text{if} \quad rand_{j,i} > CR \quad \text{or} \quad j \neq I_{rand} \end{cases} \tag{17}$$

With $rand_{j,i} \sim U(0, 1)$, I_{rand} is a random integer from $(1, 2, \dots, D)$ where D is the solution's dimension i.e. number of control variables. I_{rand} ensures that $V_{i,G+1} \neq X_{i,G}$.

Selection

The target vector $X_{i,G}$ is compared with the trial vector $V_{i,G+1}$ and the one with the better fitness value is admitted to the next generation. The selection operation in DE can be represented by the following equation:

$$X_{i,G+1} = \begin{cases} U_{i,G+1} & \text{if } f(U_{i,G+1}) < f(X_{i,G}) \\ X_{i,G} & \text{otherwise} \end{cases} \tag{18}$$

where $i \in [1, N_P]$.

RESULTS AND DISCUSSION

Application of DE

The model of the system under study has been developed in MATLAB/SIMULINK environment and DE program has been written (in .mfile). The developed model is simulated in a separate program (by .m file using initial population/controller parameters) considering a 10% step load change in area 1. The objective function is calculated in the .m file and used in the optimization algorithm. The process is repeated for each individual in the population. Using the objective function values, the population is modified by DE for the next generation. In Appendix B, the method of calculating the system eigenvalues, minimum damping ratio and settling times have been provided.

Implementation of DE requires the determination of six fundamental issues: DE step size function also called scaling factor (F), crossover probability (CR), the number of population (N_P), initialization, termination and evaluation function. The scaling factor is a value in the range (0, 2) that controls the amount of perturbation in the mutation process. Crossover probability (CR) constants are generally chosen from the interval (0.5, 1). If the parameter is co-related, then high value of CR work better, the reverse is true for no correlation [18] and [19]. DE offers several variants or strategies for optimization denoted by $DE/x/y/z$, where x = vector used to generate mutant vectors, y = number of difference vectors used in the mutation process and z = crossover scheme used in the crossover operation. In the present study, a population size of N_P = 50, generation number G = 100, step size F = 0.8 and crossover probability of CR = 0.8 have been used. The strategy employed is: DE/best/1/exp. Optimization is terminated by the prespecified number of generations for DE. One more important factor that affects the optimal solution more or less is the range for unknowns. For the very first execution of the program, a wider solution space can be given and after getting the solution one can shorten the solution space nearer to the values obtained in the previous iteration. Here the upper and lower bounds of the gains are chosen as (1, −1). The flow chart of the DE algorithm employed in the present study is given in Fig. 2. Simulations were conducted on an Intel, core 2 Duo CPU of 2.4 GHz and 2 GB MB RAM computer in the MATLAB 7.10.0.499 (R2010a) environment. The optimization was repeated 20 times and the best final solution among the 20 runs is chosen as proposed controller parameters. The best final solutions obtained in the 20 runs are shown in Table 1.

Table 1. Tuned controller parameters for different objective function.

Objective function/controller parameters	J_1	J_2	J_3
Proportional gain (K_P)	−0.2146	−0.4741	−0.4233
Integral gain (K_I)	0.4335	0.3047	0.2879

Simulation results

Table 2 shows the system eigenvalues, minimum damping ratio, settling time (2%) and various error criteria. To show the effectiveness of the proposed DE method for optimizing controller parameters, results are compared with some recently published modern heuristic optimization methods such as Bacteria Foraging Optimization Algorithm (BFOA) and Genetic Algorithm (GA) for the same interconnected power system [24]. It is clear from Table 2 that the system with conventional controller is provides small damping factor with a minimum damping ratio ($\zeta = 0.0206$) and maximum ITAE value (ITAE = 3.7568). With the proposed DE technique using ITAE as an objective function (DE-J_1), minimum ITAE value (ITAE = 0.9911) is obtained compared to the other objective functions. However, the minimum damping ratios are worse than those obtained with GA and BFOA optimized PI controller, and the settling times are inferior to BFOA even though those are better than GA. With second objective function (DE-J_2), which includes damping ratios in addition to the ITAE, minimum damping ratio has been improved ($\zeta = 0.2569$) and ITAE value (ITAE = 1.5454) compared to those with GA and BFOA techniques. But the settling times are inferior to those with BFOA. However, when third objective function is used (DE-J_3), better performance is obtained in terms of minimum damping ratio ($\zeta = 0.2361$), ITAE value (ITAE = 1.6766) and settling times compared to those with BFOA technique as presented in the literature. The above analysis shows that the system performance is greatly improved by applying the proposed controller.

Table 2. System eigenvalues, minimum damping ratio and error criteria.

Parameters		Conventional	DE-J_1	DE-J_2	DE-J_3	GA [24]	BFOA [24]
System modes		-13.7342	-13.1245	-12.9767	-13.01	-13.0081	-13.1203
		-13.6218	-13.1098	-12.916	-12.9601	-12.9575	-13.1043
		-0.086 ± 4.169	-0.4813 ± 3.15	-0.745 ± 2.8024i	-0.7037 ± 2.89	-0.7025 ± 2.8907i	-0.5605 ± 3.1716i
		-0.9568 ± 3.40	-1.0009 ± 1.88	-0.8499 ± 1.1203i	-1.0637 ± 1.32	-1.0391 ± 1.3059i	-1.1751 ± 1.9112i
		-1.8538	-0.8981 ± 0.27	-0.7083 ± 0.5092i	-0.733 ± 0.347	-0.7351 ± 0.3756i	-1.1967
		-0.2355	-0.7717	-1.2675	-0.7958	-0.8476	-0.4454
		-0.2359					-0.4288
Minimum damping ratio		0.0206	0.1509	0.2569	0.2361	0.174	0.2359
T_S (s)	Δf_1	45	8.96	6.39	5.38	10.59	5.52
	Δf_2	45	8.16	6.69	6.95	11.39	7.09
	ΔP_{Tie}	28	5.75	4.85	6.21	9.37	6.35
ITAE		3.7568	0.9911	1.5454	1.6766	2.7475	1.7975

Time domain simulations are performed for step load change at different locations and with parameter variations. The response with conventionally optimized PI controller is shown with dotted lines (with legend 'PI: CONV'), the response with PI controller optimized employing DE algorithm using objective function J_1, J_2 and J_3 are shown with dash line (with legend PI: DE-J_1), dash-dot line (with legend PI: DE-J_2) and solid line (with legend PI: DE-J_3).The following cases are considered.

Case A: Step load change in area-1
A step increase in demand of 0.1 p.u. is applied at $t = 0$ s in area-1 and the system dynamic responses are shown in Figure 3, Figure 4 and Figure 5. Critical analysis of the dynamic responses clearly reveals better dynamic performance is obtained with minimum settling time and oscillations when objective function J_3 is used.

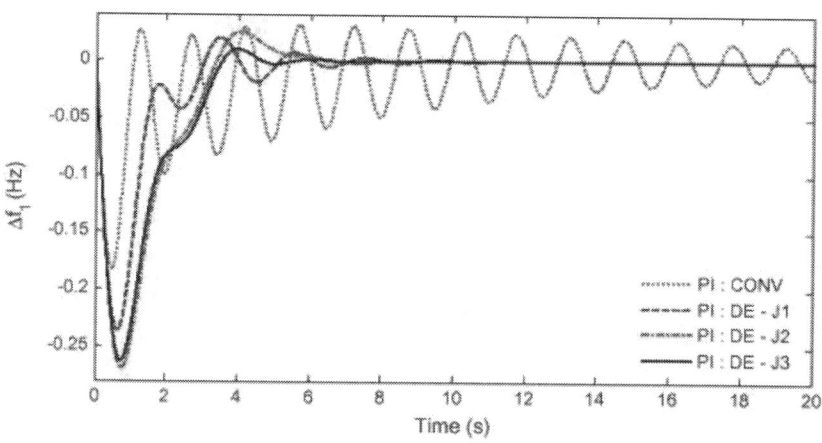

Figure 3. Change in frequency of area-1 for 0.1 p.u. change in area-1.

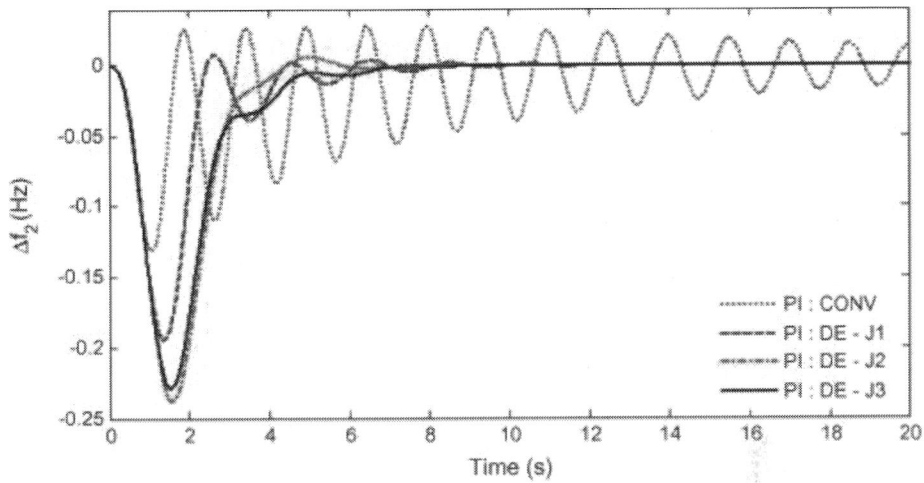

Figure 4. Change in frequency of area-2 for 0.1 p.u. change in area-1.

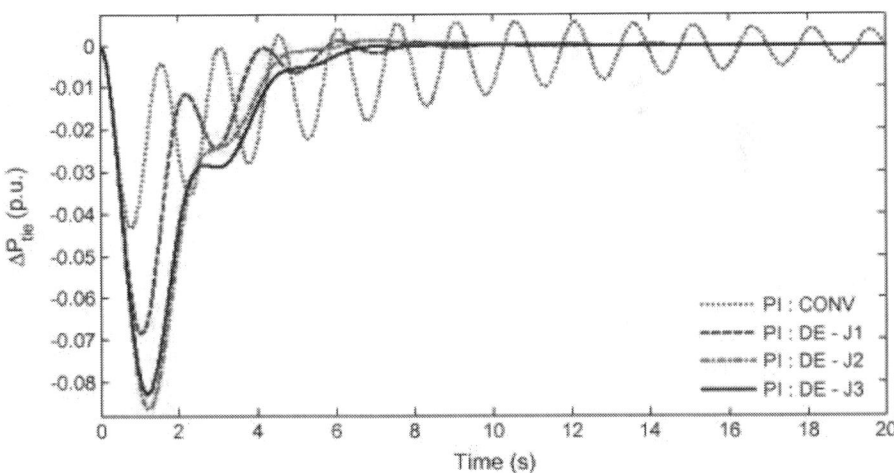

Figure 5. Change in tie line power for 0.1 p.u. change in area-1.

Case B: Step load change in area-2
Figure 6, Figure 7 and Figure 8 show the system dynamic response for a step increase in demand of 0.1 p.u. in area-2. It is clear from Figure 6, Figure 7 and Figure 8 that the designed controllers perform satisfactorily for the change in location of the disturbance

and the dynamic performance with DE-J_3out performs the other two with minimum oscillation and settling times.

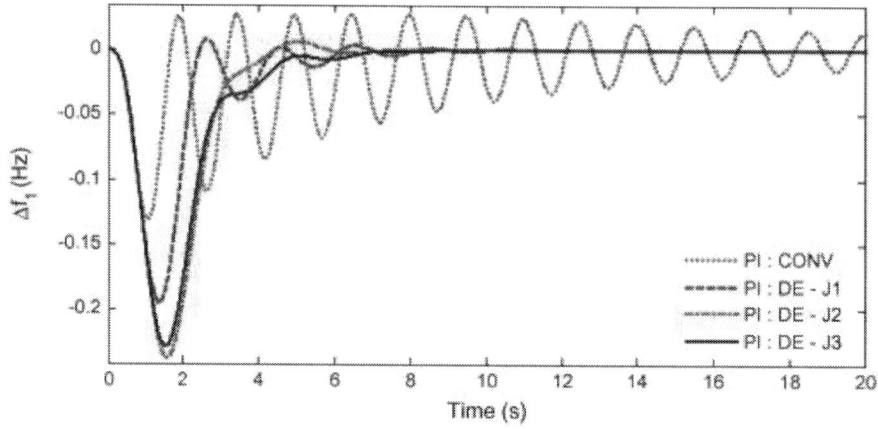

Figure 6. Change in frequency of area-1 for 0.1 p.u. change in area-2.

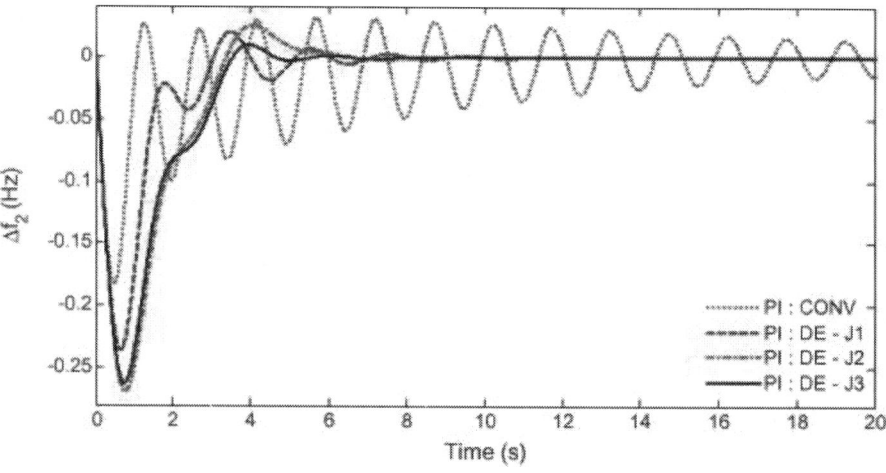

Figure 7. Change in frequency of area-2 for 0.1 p.u. change in area-2.

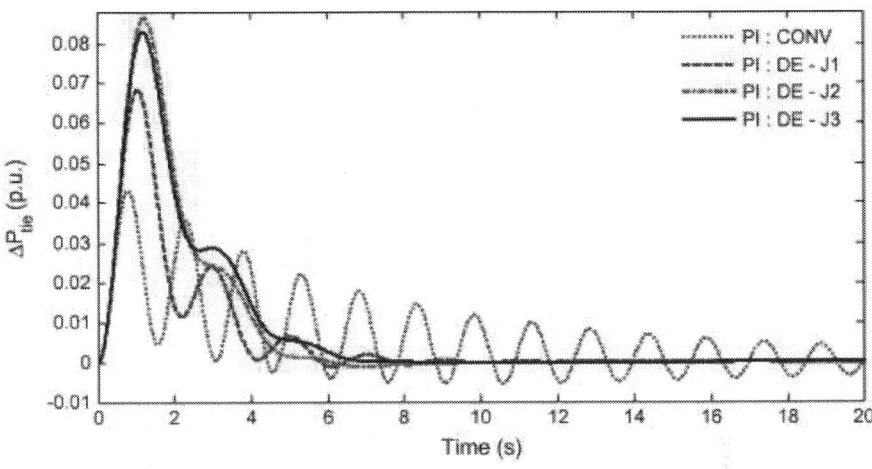

Figure 8. Change in tie line power for 0.1 p.u. change in area-2.

Case C: Step load change in both areas

A step increase in demand of 0.1 p.u. in are-1 and a step increase in demand of 0.2 p.u. in area-2 are considered simultaneously. Figure 9, Figure 10 and Figure 11 show the system dynamic response from which it is clear that the proposed controller tuned objective function J_3 achieves good dynamic performance for the power system compared to the other alternatives.

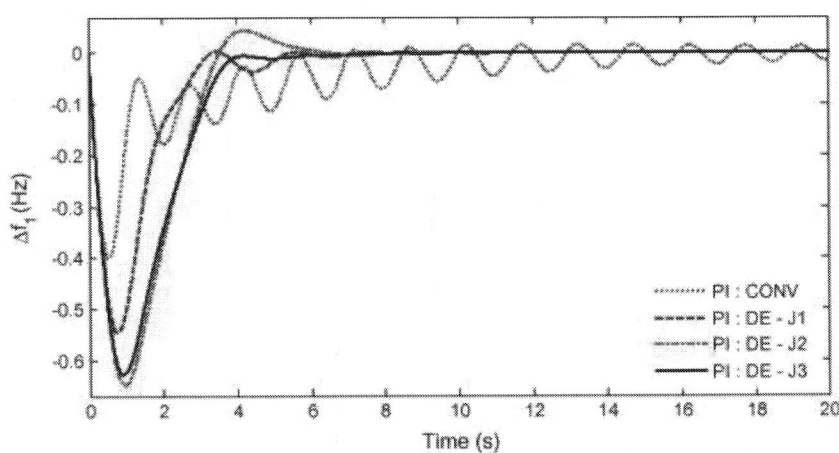

Figure 9. Change in frequency of area-1 for 0.1 p.u. change in area-1 and 0.2 p.u. change in area-2.

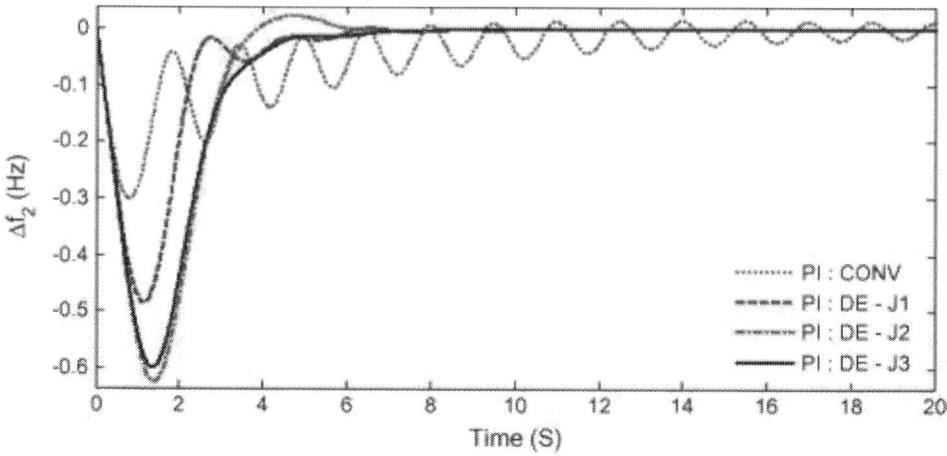

Figure 10. Change in frequency of area-2 for 0.1 p.u. change in area-1 and 0.2 p.u. change in area-2.

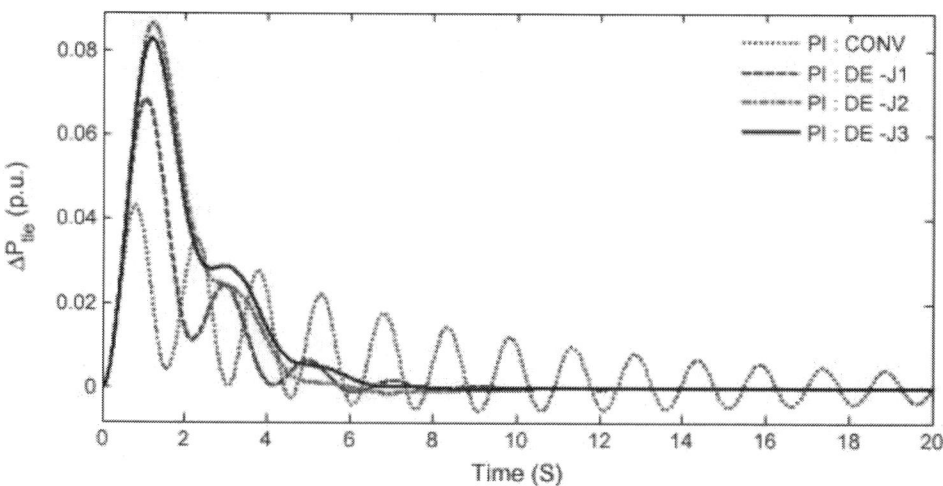

Figure 11. Change in tie line power for 0.1 p.u. change in area-1 and 0.2 p.u. change in area-2.

Case D: Sensitivity analysis

To study the robustness the system to wide changes in the operating conditions and system parameters, sensitivity analysis is carried out [23], [27], [28] and [29]. The operating load condition and time constants of speed governor, turbine, tie-line power are varied in the range of +50% to −50% from their nominal values in steps of 25% taking one at a time. Due to its superior performance, the controller parameters obtained using the objective function J_2 are considered in all cases. The results obtained are provided inTable 3. The system modes under these cases are shown in Table 4. It is obvious fromTable 3 that the system performances hardly change when the operating load condition and system parameters are changed. It is also evident from Table 4 that the eigenvalues lie in the left half of s-plane for all the cases thus maintain the stability. The frequency deviation responses for 0.1 p.u. change in area-1 with these varied conditions are shown in Figure 12, Figure 13, Figure 14 and Figure 15. It can be observed from Figure 12,Figure 13, Figure 14 and Figure 15 that there is negligible effect of the variation of operating loading conditions and system time constants on the frequency deviation responses with the controller parameters obtained at nominal values. So it can be concluded that, the proposed control strategy provides a robust and stable control satisfactorily and the optimum values of controller parameters obtained at the nominal loading with nominal parameters, need not be reset for wide changes in the system loading or system parameters.

Table 3. Sensitivity analysis.

Parameter variation	% Change	Performance index	Settling time T_s (s)			Minimum damping ratio
		ITAE	Δf_1	Δf_2	ΔP_{Tie}	
Loading condition Case-A	+25 Case-A1	1.6903	5.42	6.97	6.21	0.2387
	+50 Case-A2	1.7040	5.45	7	6.2	0.2412
	-25 Case-A3	1.6630	5.33	6.92	6.21	0.2336
	−50 Case-A4	1.6494	5.27	6.91	6.22	0.231
T_G Case-B	+25 Case-B1	1.6577	5.39	6.89	6.14	0.2192
	+50 Case-B2	1.6418	7.35	6.86	6.08	0.2024
	−25 Case-B3	1.6961	5.37	7.01	6.31	0.2526
	−50 Case-B4	1.7158	5.33	7.05	6.4	0.2676
T_T Case-C	+25 Case-C1	1.6254	7.63	7.02	6.16	0.1872
	+50 Case-C2	1.7533	8.89	9.86	9.22	0.1543
	−25 Case-C3	1.7505	5.22	6.91	6.5	0.3139
	−50 Case-C4	1.8248	5.25	6.78	7.01	0.4258
T_{12} Case-D	+25 Case-D1	1.6629	6.82	7.53	5.84	0.2095
	+50 Case-D2	1.6538	6.43	7.32	5.52	0.1895
	−25 Case-D3	1.6997	5.22	7.53	6.61	0.274
	−50 Case-D4	1.7463	6.34	8.18	6.91	0.3333

Table 4. System modes for each case.

Cases A	System modes	Cases-B	System modes	Cases-C	System modes	Cases-D	System modes
	−13.0085		−10.6599		−12.8884		−13.02
	−12.9579		−10.589		−12.8487		−12.9575
	−0.7110 ± 2.8927i		−0.6576 ± 2.9275i		−0.5526 ± 2.8998i		−0.6820 ± 3.1831i
Case-A1	−1.0523 ± 1.3136i	Case-B1	−0.9813 ± 1.3475i	Case-C1	−0.7910 ± 1.3409i	Case-D1	−1.0391 ± 1.3059i
	−0.7326 ± 0.3780i		−0.7041 ± 0.3979i		−0.6115 ± 0.4300i		−0.7497 ± 0.3857i
	−0.8333		−0.8317		−0.7858		−0.8476
	−13.009		−9.1448		−12.8143		−13.0316
	−12.9583		−9.0554		−12.7817		−12.9575
	−0.7195 ± 2.8947i		−0.6108 ± 2.9549i		−0.4506 ± 2.8863i		−0.6660 ± 3.4507i
Case-A2	−1.0653 ± 1.3214i	Case-B2	−0.9220 ± 1.3803i	Case-C2	−0.6231 ± 1.3249i	Case-D2	−1.0391 ± 1.3059i
	−0.7302 ± 0.3804i		−0.6751 ± 0.4152i		−0.5284 ± 0.4462i		−0.7598 ± 0.3922i
	−0.8194		−0.8174		−0.7443		−0.8476
	−13.0077		−17.028		−13.2326		−12.9959
	−12.9571		−16.9971		−13.1632		−12.9575
	−0.6940 ± 2.8888i		−0.7426 ± 2.8448i		−0.9269 ± 2.8038i		−0.7312 ± 2.5659i
Case-A3	−1.0257 ± 1.2983i	Case-B3	−1.0936 ± 1.2554i	Case-C3	−1.4312 ± 1.0722i	Case-D3	−1.0391 ± 1.3059i
	−0.7376 ± 0.3732i		−0.7684 ± 0.3467i		−0.8935		−0.7125 ± 0.3582i
	−0.8624		−0.8656		−1.0145		−0.8476
					−0.9689		
	−13.0072		−25.2251		−13.7968		−12.9834
	−12.9567		−25.2107		−13.6893		−12.9575
	−0.6855 ± 2.8868i		−0.7752 ± 2.7908i		−1.1523 ± 2.4484i		−0.7767 ± 2.1969i
Case-A4	−1.0121 ± 1.2908i	Case-B4	−1.1433 ± 1.1964i	Case-C4	−3.3517	Case-D4	−1.0391 ± 1.3059i
	−0.7401 ± 0.3707i		−0.8039 ± 0.3084i		−2.492		−0.6733 ± 0.3212i
	−0.8776		−0.8861		−1.0879 ± 0.3872i		−0.8476
					−0.6232		

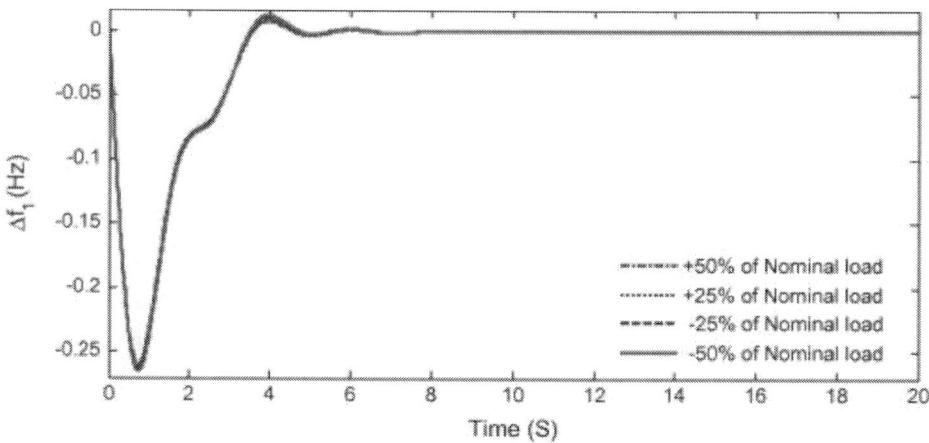

Figure 12. Change in frequency of area-1 for 0.1 p.u. change in area-1with change in loading.

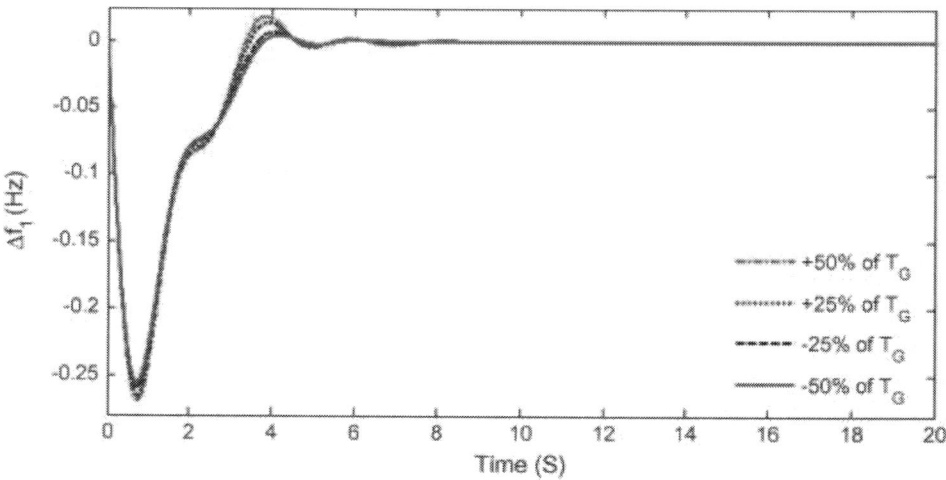

Figure 13. Change in frequency of area-1 for 0.1 p.u. change in area-1with change in T_G.

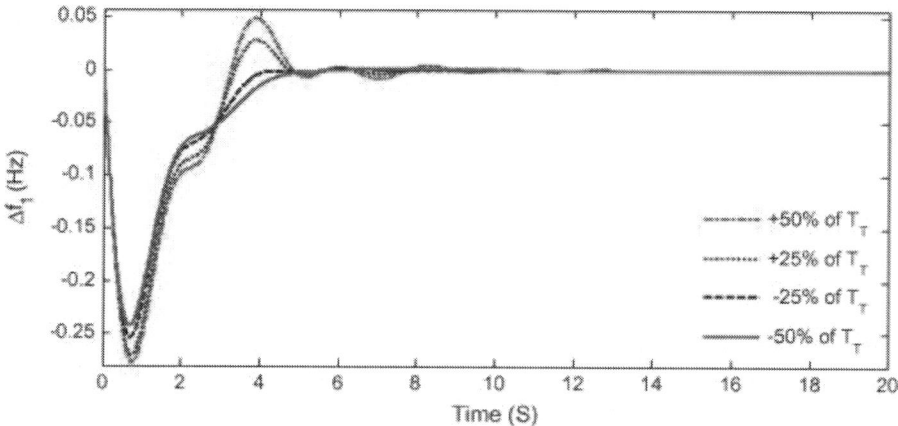

Figure 14. Change in frequency of area-1 for 0.1 p.u. change in area-1with change in T_T.

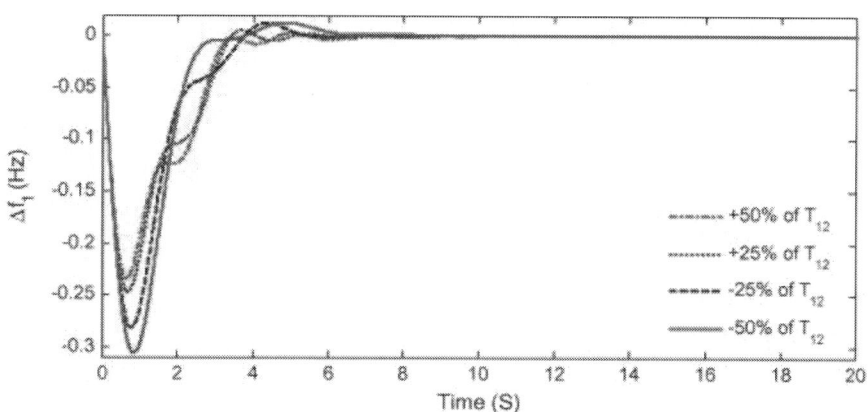

Figure 15. Change in frequency of area-1 for 0.1 p.u. change in area-1with change in T_{12}.

CONCLUSION

This study presents the design and performance evaluation of Differential Evolution (DE) optimized Proportional-Integral (PI) controller for Automatic Generation Control (AGC) of an interconnected power system. For the optimization of controller parameters using modern heuristic optimizations techniques, selection of suitable objective function is very important. In view of the above, different objective functions using Time multiply Absolute Error (ITAE), damping ratio of dominant eigenvalues and settling time with appropriate weight coefficients are employed to increase the performance of the controller. The results obtained from the simulations show that the proposed control strategy optimized with new objective function achieves better dynamic performances than the standard objective functions. The superiority of the proposed design approach has been shown by comparing the results with some recently published modern heuristic optimization techniques such as Bacteria Foraging Optimization Algorithm (BFOA) and Genetic Algorithm (GA) based PI controller for the same interconnected power system. Further, robustness analysis is carried out which demonstrates the robustness of the proposed DE optimized PI controller to wide changes in loading condition and system parameters.

APPENDIX A

Nominal parameters of the system investigated are:

P_R = 2000 MW (rating), P_L = 1000 MW (nominal loading); f = 60 Hz, B_1, B_2 = 0.045 p.u. MW/Hz; R_1 = R_2 = 2.4 Hz/p.u.; T_{G1} = T_{G2} = 0.08 s; T_{T1} = T_{T2} = 0.3 s; K_{PS1} = K_{PS2} = 120 Hz/p.u. MW; T_{PS1} = T_{PS2} = 20 s; T_{12} = 0.545 p.u.; a_{12} = −1.

APPENDIX B

It is desirable that the transient response of a system be sufficiently fast with small settling time and be adequately damped. The settling time is the time required for the response curve to reach and stay within a range about the final value of size specified by absolute percentage of the final value (usually 2% or 5%). To make the settling time small, the damping ratio should not be too small.

The MATLAB programme to find out system eigenvalues, settling time and minimum damping ratio is given below:

[A, B, C, D] = linmod('Model');	% Model is the SIMULINK model of system
Eigen_Values = eig(A)	% Computes the system eigenvalues
[wn, Z] = damp(A);	% Computes the natural frequencies and damping factors
Minim_Damping_Ratio = min(abs(Z))	% Computes minimum damping ratio

sim('Model', 50);
time=[0: 0.01: 50];
for t = 1: 5001	

if (Del_f_1(t) ⩾ 0.002 || Del_f_1(t) ⩽ −0.002)	
st = t; end end Tettling_Time_for_Delf_1 = time(st)	% Computes the settling time

REFERENCES

1. Kundur P. Power system stability and control, TMH, 8th reprint; 2009.
2. Elgerd OI. Electric energy systems theory. An introduction. New Delhi: Tata McGraw-Hill; 1983.

3. Ibraheem, Kumar P, Kothari DP. Recent philosophies of automatic generation control strategies in power systems. IEEE Trans Power Syst 2005;20(1):346–57.

4. Kothari ML, Nanda J, Kothari DP, Das D. Discrete-mode automatic generation control of a two-area reheat thermal system with new area control error. IEEE Trans Power Syst 1989;4(2):730–8.

5. Shoults RR, Jativa Ibarra JA. Multi area adaptive LFC developed for a comprehensive AGC simulation. IEEE Trans Power Syst 1993;8(2):541–7.

6. Zeynelgil HL, Demiroren A, Sengor NS. The application of ANN technique to automatic generation control for multi-area power system. Electr Power Energy Syst 2002;24(5):345–54.

7. Demiroren A, Zeynelgil HL, Sengor NS. Application of ANN technique to load frequency control for three area power system. In: IEEE power technol conf, Porto, vol. 2; 2001.

8. Wu QH, Hogg BW, Irwin GW. A neural network regulator for turbo generator. IEEE Trans Neural Netw 1992;3(1):95–100.

9. Chaturvedi DK, Satsangi PS, Kalra PK. Load frequency control: a generalized neural network approach. Electr Power Energy Syst 1999;21(6):405–15.

10. Ghosal SP. Optimization of PID gains by particle swarm optimization in fuzzy based automatic generation control. Electr Power Syst Res 2004;72(3):203–12.

11. Ghosal SP. Application of GA/GA-SA based fuzzy automatic generation control of a multi-area thermal generating system. Electr Power Syst Res 2004;70(2):115–27.

12. Talaq J, Al-Basri F. Adaptive fuzzy gain scheduling for load frequency control. IEEE Trans Power Syst 1999;14(1):145–50.

13. Imthias Ahamed TP, Nagendra Rao PS, Sastry PS. A reinforcement learning approach to automatic generation control. Electr Power Syst Res 2002;63(1):9–26.

14. Hosseini SH, Etemadi AH. Adaptive neuro-fuzzy inference system based automatic generation control. Electr Power Syst Res 2008;78(7):1230–9.

15. Khuntia SR, Panda S. Simulation study for automatic generation control of a multi-area power system by ANFIS approach. Appl Soft Comput 2012;12(1):333–41.

16. Panda S, Padhy NP. Comparison of particle swarm optimization and genetic algorithm for FACTS-based controller design. Appl Soft Comput 2008;8(4):1418–27.

17. Panda S, Padhy NP. Optimal location and controller design of STATCOM for power system stability improvement using PSO. J Franklin Inst 2008;345(2):166–81.

18. Panda S. Differential evolution algorithm for SSSC-based damping controller design considering time delay. J Franklin Inst 2011;348(8):1903–26.

19. Panda Sidhartha. Robust coordinated design of multiple and multi-type damping controller using differential evolution algorithm. Electr Power Energy Syst 2011;33:1018–30.

20. Panda S. Multi-objective evolutionary algorithm for SSSC-based controller design. Electr Power Syst Res 2009;79:937–44.

21. Panda S. Application of non-dominated sorting genetic algorithm-II technique for optimal FACTS-based controller design. J Franklin Inst 2010;347(7):1047–64.

22. Panda S. Multi-objective PID controller tuning for a FACTSbased damping stabilizer using non-dominated sorting genetic algorithm-II. Int J Electr Power Energy Syst 2011;33:1296–308.

23. Nanda J, Mishra S, Saikia LC. Maiden application of bacterial foraging based optimization technique in multiarea automatic generation control. IEEE Trans Power Syst 2009;24(2):602–9.

24. Ali ES, Abd-Elazim SM. Bacteria foraging optimization algorithm based load frequency controller for interconnected power system. Electr Power Energy Syst 2011;33:633–8.

25. Stron R, Price K. Differential evolution – a simple and efficient adaptive scheme for global optimization over continuous spaces. J Glob Optim 1995;11:341–59.

26. Ogatta K. Modern control engineering. NJ, USA: Prentice Hall; 1990.

27. Saikia LC et al.. Performance comparison of several classical controllers in AGC for multi-area interconnected thermal system. Int J Electr Power Energy Syst 2011;33:394–401.
28. Gozde H et al.. Comparative performance analysis of artificial bee colony algorithm in automatic generation control for interconnected reheat thermal power system. Int J Electr Power Energy Syst 2012;42:167–78.
29. Parmar KPS et al.. Load frequency control of a realistic power system with multi-source power generation. Int J Electr Power Energy Syst 2012;42:426–33.

CITATION

Umesh Kumar Rout, Rabindra Kumar Sahu, Sidhartha Panda, Design and analysis of differential evolution algorithm based automatic generation control for interconnected power system, Ain Shams Engineering Journal, Volume 4, Issue 3, September 2013, Pages 409-421, ISSN 2090-4479, http://dx.doi.org/10.1016/j.asej.2012.10.010.

CHAPTER 6

Automatic Generation Control With Thyristor Controlled Series Compensator Including Superconducting Magnetic Energy Storage Units

Saroj Padhan, Rabindra Kumar Sahu, Sidhartha Panda

Department of Electrical Engineering, Veer Surendra Sai University of Technology (VSSUT), Burla 768018, Odisha, India

ABSTRACT

In the present work, an attempt has been made to understand the dynamic performance of Automatic Generation Control (AGC) of multi-area multi-units thermal–thermal power system with the consideration of Reheat turbine, Generation Rate Constraint (GRC) and Time delay. Initially, the gains of the fuzzy PID controller are optimized using Differential Evolution (DE) algorithm. The superiority of DE is demonstrated by comparing the results with Genetic Algorithm (GA). After that performance of Thyristor Controlled Series Compensator (TCSC) has been investigated. Further, a TCSC is placed in the tie-line and Superconducting Magnetic Energy Storage (SMES) units are considered in both areas. Finally, sensitivity analysis is performed by varying the system parameters and operating load conditions from their nominal values. It is observed that the optimum gains of the proposed controller need not be reset even if the system is subjected to wide variation in loading condition and system parameters.

INTRODUCTION

Load Frequency Control (LFC) is a very important issue in modern power system operation and control for supplying sufficient and reliable electric power with good quality. The main goal of the LFC is to maintain the system frequency of each area and the tie line power within tolerable limits with variation in load demands [1]. For power balance, the power generated should match with the total load demanded and associated system losses. However the load demands fluctuate randomly causing a mismatch in the power balance and thereby deviations in the area frequencies and tie-line powers from their respective scheduled values, called Automatic Load Frequency Control (ALFC) [2] and [3]. Due to the complexity of the modern power system, superior intelligent control design is essential. Literature study reveals that several control strategies have been proposed by many researchers over the past decades for LFC of power system. Many control and optimization techniques such as classical, optimal, Genetic Algorithm (GA), Particle Swarm Optimization (PSO), Fuzzy Logic Controller (FLC), and Artificial Neural Network (ANN), have been proposed for LFC [4], [5], [6], [7],[8] and [9]. Design of a controller for AGC can be divided into two groups. In the 1st group the controller gains are tuned by a suitable optimization algorithm. In the 2nd group researchers have adopted self-tuning techniques with the help of neural network and fuzzy logic. Fuzzy logic controllers have been successfully used for analysis and control of non-linear system in the past decades. Yesil et al. [10] have used a self-tuning fuzzy PID type controller for load frequency control of a two-area interconnected system. Khuntia and Panda [11] have used ANFIS approach for AGC of a three area system. Ghosal [8] have used PSO optimization technique to optimize the PID controller gain for a fuzzy based LFC. These methods provide good performances but the transient responses are oscillatory in nature. Fuzzy logic based PID controller can be successfully used for all non-linear system but there is no specific mathematical formulation to decide the proper choice of fuzzy parameters (such as inputs, scaling factors, membership functions, and rule base). Normally these parameters are selected by using certain empirical rules and therefore may not be the optimal parameters. Improper selection of input–output scaling factor may affect the performance of FLC to a greater extent.

To get an accurate insight into the AGC problem, it is necessary to include the important physical constraints in the system model. The major physical constraints that affect the power system performance are Generation Rate Constraint (GRC) and time delay. The Flexible AC Transmission System (FACTS) controllers [12] play a crucial role to enhance power system stability in addition to control the power flow in an interconnected power system. Several studies have explored the potential of using FACTS devices for better power system control since it provides more flexibility. A Superconducting Magnetic Energy Storage (SMES) is capable of controlling both active and reactive power simultaneously. SMES unit with small storage capacity can be essential not only as a fast energy compensation device for power consumptions of large loads, but also as a stabilizer of frequency oscillations [13]. TCSC is one of the FACTS controller which is enhanced the power system dynamics, power transfer capability of transmission lines and dynamic stability [14].

It obvious from the literature survey that the performance of the power system not only depends on the controller structure but also depends on the artificial optimization technique. Hence, proposing and implementing new high performance heuristic optimization algorithms to real world problems are always welcome. Differential Evolution (DE) is a population-based direct search algorithm for global optimization capable of handling non-differentiable, non-linear and multi-modal objective functions, with few, easily chosen, control parameters [15] and [16]. However, the success of DE in solving a specific problem crucially depends on appropriately choosing trial vector generation strategies and their associated control parameter values namely the step size F, crossover probability CR, number of population NP and generations G [17].

In view of the above, a Differential Evolution (DE) optimized fuzzy PID controller is proposed for Load Frequency Control (LFC) of multi-area multi-units thermal–thermal power system with the consideration of reheat turbine, Generation Rate Constraint (GRC) and time delay. The superiority of the proposed approach is shown by comparing the results with GA for the same power system. Further, TCSC is employed in series with the tie-line in coordination with SMES to improve the dynamic performance of the power system. Finally, sensitivity analysis is carried out by varying the loading condition and system parameters.

MATERIALS AND METHODS

System under study

The system under investigation consists of two area interconnected thermal power system as shown in Fig. 1. Area 1 comprises two reheat thermal power units. Area 2 comprises two non-reheat thermal units. In Fig. 1, B_1 and B_2 are the frequency bias parameters; ACE_1 and ACE_2 are area control errors; R_1, R_2 and R_3, R_4 are the governor speed regulation parameters in pu Hz for area 1 and area 2 respectively; T_{G1}, T_{G2} and T_{G3}, T_{G4} are the speed governor time constants in sec for area 1 and area 2 respectively; T_{T1}, T_{T2} and T_{T3}, T_{T4} are the turbine time constant in sec for area 1 and area 2 respectively; ΔP_{D1} and ΔP_{D2} are the load demand changes; ΔP_{Tie} is the incremental change in tie line power (p.u); K_{Ps1} and K_{Ps2} are the power system gains; T_{Ps1} and T_{Ps2} are the power system time constant in sec; T_{12} is the synchronizing coefficient and ΔF_1 and ΔF_2 are the system frequency deviations in Hz. To get an accurate insight into the AGC problem, it is essential to include the important inherent requirement and the basic physical constraints and include them model. The important constraints that affect the power system performance are Generation Rate Constraint (GRC), and Time delay. In view of the above, the effect of GRC and Time delay are included to a power system model. Time delays can degrade a system's performance and even cause system instability. In a power system having steam plants, power generation can change only at a specified maximum rate. In thermal power plants, power generation can change only at a specified maximum/minimum rate known as Generation Rate Constraint (GRC). In the present study, a GRC of 3%/min for reheat and 10%/ min for non-reheat thermal units are considered [18] and [19]. Also in the present study, a time delay of 50 ms is considered[20]. The relevant parameters are given in Appendix A.

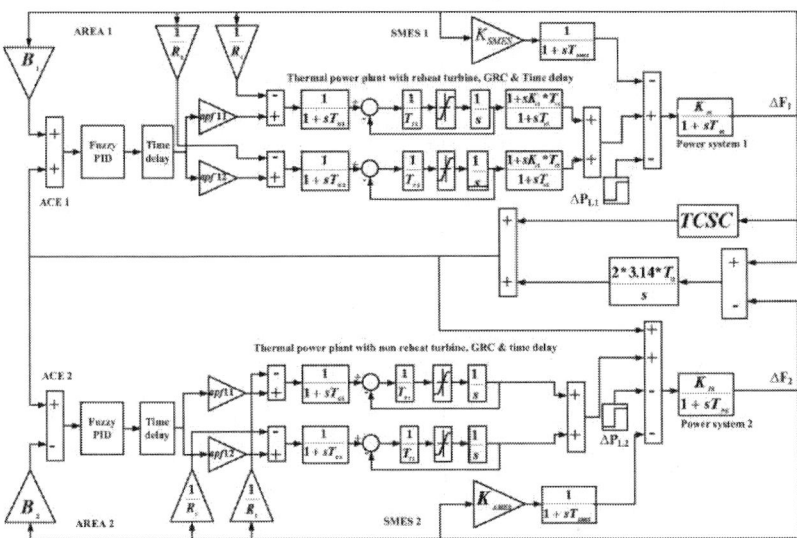

Figure 1. MATLAB/SIMULINK model of multi-area multi-units thermal system.

Control structure and objective function

To control the frequency, fuzzy PID controllers are provided in each area. The structure of fuzzy PID controller is shown in Fig. 2[21] and [22].

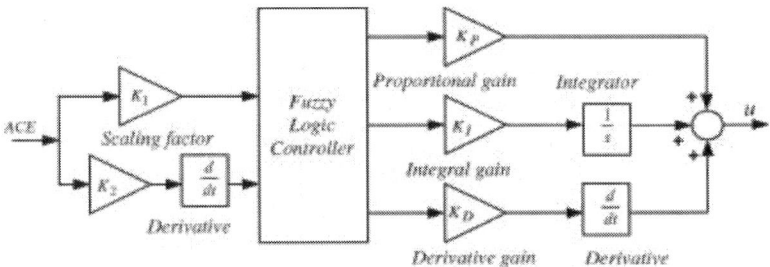

Figure 2. Structure of proposed fuzzy PID controller.

The error inputs to the controllers are the respective area control errors (ACE) given by:

$$e_1(t)=ACE_1=B_1\Delta F_1+\Delta P_{Tie} \tag{1}$$

$$e_2(t)=ACE_2=B_2\Delta F_2-\Delta P_{Tie} \tag{2}$$

Fuzzy controller uses error (e) and derivative of error $(\dot{e})$ as input signals. The outputs of the fuzzy controllers u_1 and u_2 are the control inputs of the power system i.e. the reference power settings ΔP_{ref1} and ΔP_{ref2}. The input scaling factors are the tuneable parameters K_1 and K_2. The proportional, integral and derivative gains of fuzzy controller are represented by K_P, K_I and K_D respectively. Triangular membership functions are used with five fuzzy linguistic variables such as NB (negative big), NS (negative small), Z (zero), PS (positive small) and PB (positive big) for both the inputs and the output. Membership functions for error, error derivative and FLC output are shown in Fig. 3. Mamdani fuzzy interface engine is selected for this work. The FLC output is determined by using center of gravity method of defuzzification. The two-dimensional rule base for error, error derivative and FLC output is shown in Table 1.

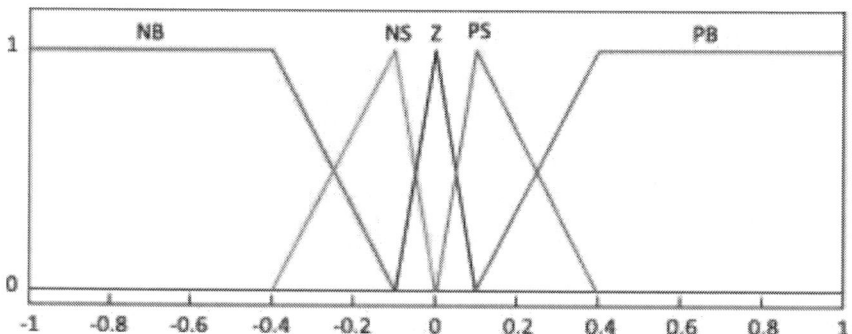

Figure 3. Membership functions for error, error derivative and FLC output.

Table 1. Rule base for error, derivative of error and FLC output.

e	$\dot{e}$				
	NB	NS	Z	PS	PB
NB	NB	NB	NS	NS	Z
NS	NB	NS	NS	Z	PS
Z	NS	NS	Z	PS	PS
PS	NS	Z	PS	PS	PB
PB	Z	PS	PS	PB	PB

In the design of modern heuristic optimization technique based controller, the objective function is first defined based on the desired specifications and constraints. Typical output specifications in the time domain are peak overshooting, rise time, settling time, and steady-state error. It has been reported in the literature that Integral of Time multiplied Absolute Error (ITAE) gives a better performance compared to other integral based performance criteria [23]. Therefore in this paper ITAE is used as objective function to optimize the scaling factors and proportional, integral and derivative gains of fuzzy PID controller. Expression for the ITAE objective function is depicted in Eq. (3).

$$J = ITAE = \int_0^{t_{sim}} \left(|\Delta F_1| + |\Delta F_2| + |\Delta P_{Tie}| \right) \cdot t \cdot dt$$

(3)

In the above equation, ΔF_1 and ΔF_2 are the system frequency deviations; ΔP_{Tie} is the incremental change in tie line power; t_{sim} is the time range of simulation.

Modeling of TCSC in AGC

It is well known that the reactance adjusting of Thyristor Controlled Series Compensator (TCSC) is a complex dynamic process. Effective design and accurate evaluation of the TCSC control strategy depends on the simulation accuracy of this process. Basically a TCSC consists of three components: capacitor banks, bypass inductor and bidirectional thyristors. The firing angles of the thyristors are controlled to adjust the TCSC reactance in accordance with a system control algorithm, normally in response to some system parameter variations. According to the variation in the thyristor firing angle, this process can be modeled as a fast switch between corresponding reactance offered to the power system. Both capacitive and inductive reactance compensation are possible by proper selection of capacitor and inductor values of the TCSC device. TCSC is considered as a variable reactance, the value of which is adjusted automatically to constrain the power flow across the branch to a specified value. The variable reactance X_{TCSC} represents the net equivalent reactance of the TCSC, when operating in either the inductive or the capacitive mode [14]. Fig. 4 shows the schematic diagram of a two area interconnected thermal-thermal power system with TCSC connected in series with the tie-line. For analysis, it is assumed that TCSC is connected near to the area 1. Resistance of the tie-line is neglected, since the effect on the dynamic performance is negligible. Further, the reactance to resistance ratio in a practically interconnected power system is quite high. The incremental tie-line power flow without TCSC is given by (4).

$$\Delta P_{Tie12}(s) = \frac{2\pi T_{12}^{0}}{s}[\Delta F_1(s) - \Delta F_2(s)] \qquad \text{equation(4)}$$

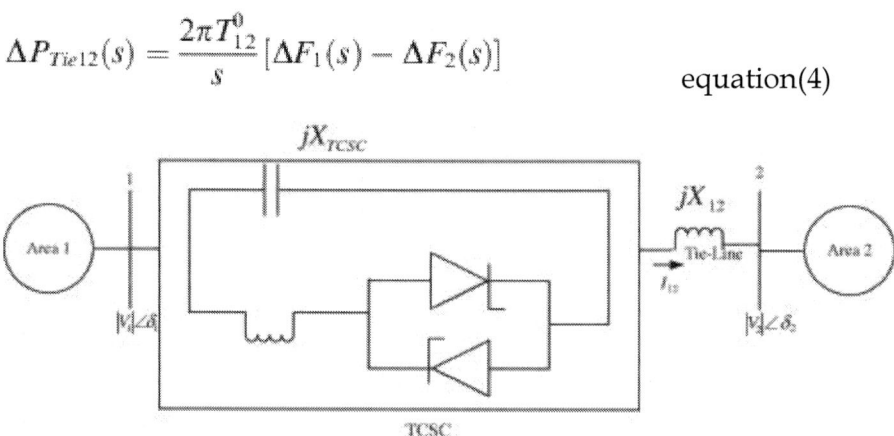

Figure 4. Two-area interconnected power system with TCSC.

In the above equation, ΔF_1 and ΔF_2 are the system frequency deviations; T^0_{12} is the synchronizing coefficient without TCSC. The line current flow from area-1 to area-2 can be written as, when TCSC is connected in series with the tie-line

$$I_{12} = \frac{|V_1|\angle(\delta_1) - |V_2|\angle(\delta_2)}{j(X_{12} - X_{TCSC})} \tag{5}$$

where X_{12} and X_{TCSC} are the tie-line and TCSC reactance respectively.

It is clear from Fig. 4 that, the complex tie-line power as

$$P_{Tie12} - jQ_{Tie12} = V_1^*I_{12} = |V_1|\angle(-\delta_1)\left[\frac{|V_1|\angle(\delta_1) - |V_2|\angle(\delta_2)}{j(X_{12} - X_{TCSC})}\right] \tag{6}$$

Solving the above equation, the real part,

$$P_{Tie12} = \frac{|V_1||V_2|}{(X_{12} - X_{TCSC})}\sin(\delta_1 - \delta_2) \tag{7}$$

The tie-line power flow can be represented in terms of % compensation (k_c) as

$$P_{Tie12} = \frac{|V_1||V_2|}{X_{12}(1 - k_c)}\sin(\delta_1 - \delta_2) \tag{8}$$

where $k_c = \frac{X_{TCSC}}{X_{12}}$, percentage of compensation offered by the TCSC

In order to obtain the linear incremental model, Eq. (8) can be rewritten as

$$\Delta P_{Tie12} = \frac{|V_1||V_2|}{X_{12}(1 - k_C^0)^2} \sin(\delta_1^0 - \delta_2^0)\Delta k_C + \frac{|V_1||V_2|}{X_{12}(1 - k_C^0)} \cos(\delta_1^0 - \delta_2^0)(\Delta\delta_1 - \Delta\delta_2) \tag{9}$$

If $f_{12}^0 = \frac{|V_1||V_2|}{X_{12}} \sin(\delta_1^0 - \delta_2^0)$ and $T_{12}^0 = \frac{|V_1||V_2|}{X_{12}} \cos(\delta_1^0 - \delta_2^0)$, then Eq. (9) is expressed as

$$\Delta P_{Tie12} = \frac{f_{12}^0}{(1 - k_C^0)^2} \Delta k_C + \frac{T_{12}^0}{(1 - k_C^0)}(\Delta\delta_1 - \Delta\delta_2) \tag{10}$$

Since $\Delta\delta_1 = 2\pi \int \Delta F_1 dt$ and $\Delta\delta_2 = 2\pi \int \Delta F_2 dt$

Taking Laplace transforms of Eq. (10) and expressed as given by (11)

$$\Delta P_{Tie12}(s) = \frac{f_{12}^0}{(1 - k_C^0)^2} \Delta k_C(s) + \frac{2\pi T_{12}^0}{s(1 - k_C^0)}[\Delta F_1(s) - \Delta F_2(s)] \tag{11}$$

From Eq. (11), the tie-line power flow can be regulated by controlling $\Delta k_c(s)$. If the control input signal to TCSC damping controller is assumed to be $\Delta Error(s)$ and the transfer function of the signal conditioning circuit is $k_c = \frac{K_{TCSC}}{1 + sT_{TCSC}}$, The expression is given (12)

$$\Delta k_C(s) = \frac{K_{TCSC}}{1 + sT_{TCSC}} \Delta Error(s) \tag{12}$$

where K_{TCSC} and T_{TCSC} is the gain and time constant of the TCSC controller respectively. As TCSC is kept near to area-1, frequency deviation ΔF_1 may be suitably used as the control signal $\Delta Error(s)$, to the TCSC unit to control the percentage incremental change in the system compensation level. Therefore,

$$\Delta k_C(s) = \frac{K_{TCSC}}{1 + sT_{TCSC}} \Delta F_1(s) \tag{13}$$

$$\Delta P_{Tie12} = \frac{2\pi T_{12}^0}{s(1 - k_C^0)} [\Delta F_1(s) - \Delta F_2(s)] + \left[\frac{f_{12}^0}{(1 - k_C^0)^2} \right] \frac{K_{TCSC}}{1 + sT_{TCSC}} \Delta F_1(s) \tag{14}$$

Modeling of SMES in AGC

Superconducting Magnetic Energy Storage (SMES) is a device which can store the electrical power from the grid in the magnetic field of a coil. The magnetic field of coil is made of superconducting wire with near-zero loss of energy. SMESs can store and refurbish huge values of energy almost instantaneously. Therefore the power system can discharge high levels of power within a fraction of a cycle to avoid a rapid loss in the line power. The SMES is consisting of inductor-converter unit, dc superconducting inductor, AC/ DC converter and a step down transformer [24]. The stability of a SMES unit is superior to other power storage devices, because all parts of a SMES unit are static. Fig. 5 shows the schematic diagram of SMES unit in the power system [13]. During normal operation of the grid, the superconducting coil will be charged to a set value (normally less than the maximum charge) from the utility grid. After charged, the superconducting coil conducts current, which supports an electromagnetic field, with virtually no losses. The coil is kept at very low temperature by immersion in a bath of liquid helium.

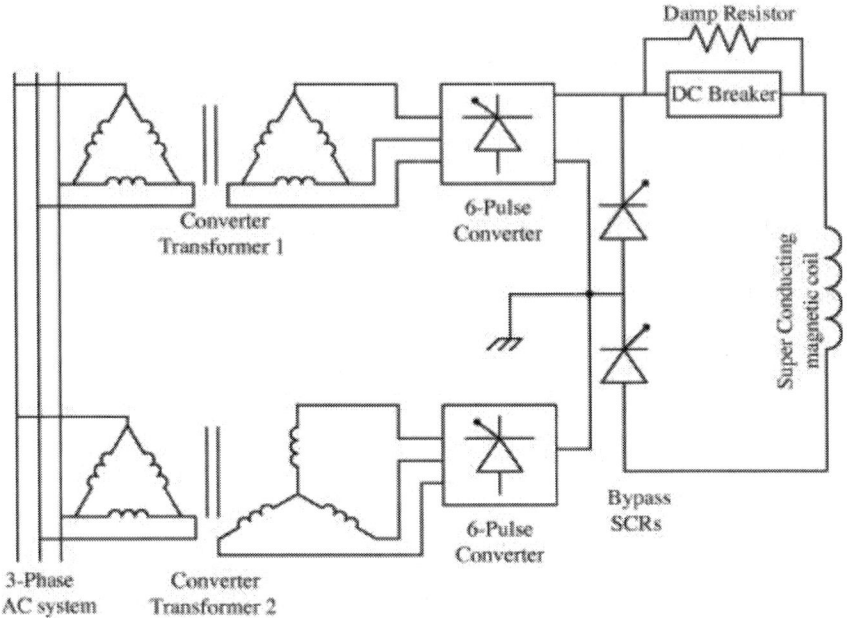

Figure 5. SMES circuit diagram.

In the present work two SMES units are established in area1 and area2 in order to stabilize frequency oscillations as shown in Fig. 1. The input signal of the SMES controller is p.u. frequency deviation (ΔF) and the output is the change in control vector [ΔP_{SMES}]. The controller gains K_{SMES} and the time constant T_{SMES} values are 0.12 and 0.03 s respectively [24].

Over view of differential evolution

Differential Evolution (DE) algorithm is a search heuristic algorithm introduced by Storn and Price [15]. It is a simple, efficient, reliable algorithm with easy coding. The main advantage of DE over Genetic Algorithm (GA) is that GA uses crossover operator for evolution while DE relies on mutation operation. The mutation operation in DE is based on the difference in randomly sampled pairs of solutions in the population. An optimization task consisting of D variables can be represented by a D-dimensional vector. A population of N_P solution vectors is randomly initialized within the parameter bounds at the beginning. The population is modified by applying mutation, crossover and selection operators. DE algorithm uses two generations; old generation and new generation of the same population size. Individuals of the current population become target vectors for the next generation. The mutation operation produces a mutant vector for each target vector, by adding the weighted difference between two randomly chosen vectors to a third vector. A trial vector is generated by the crossover operation by mixing the parameters of the mutant vector with those of the target vector. The trial vector substitutes the target vector in the next generation if it obtains a better fitness value than the target vector. The evolutionary operators are described below [25] and [26]:

Initialization of parameter

DE begins with a randomly initiated population of size N_P of D dimensional real-valued parameter vectors. Each parameter j lies within a range and the initial population should spread over this range as much as possible by uniformly randomizing individuals within the search space constrained by the prescribed lower bound X_j^L and upper bound X_j^U.

Mutation operation

For the mutation operation, a parent vector from the current generation is selected (known as target vector), a mutant vector is obtained by the differential mutation operation (known as donor vector) and finally an offspring is produced by combining the donor with the target vector (known as trial vector). Mathematically it can be expressed as:

$$V_{i,G+1} = X_{r1,G} + F \cdot (X_{r2,G} - X_{r3,G}) \tag{15}$$

where $X_{i,G}$ is the given parameter vector, $X_{r1,G}$ $X_{r2,G}$ $X_{r3,G}$ are randomly selected vector with distinct indices $i, r1, r2$ and $r3$, $V_{i,G+1}$ is the donor vector and F is a constant from $(0, 2)$

Crossover operation

After generating the donor vector through mutation the crossover operation is employed to enhance the potential diversity of the population. For crossover operation three parents are selected and the child is obtained by means of perturbation of one of them. In crossover operation a trial vector $U_{i,G+1}$ is obtained from target vector ($X_{i,G}$) and donor vector ($V_{i,G}$). The donor vector enters the trial vector with probability CR given by:

$$U_{j,i,G+1} = \begin{cases} V_{j,i,G+1} & \text{if } rand_{j,i} \leqslant CR \quad or \quad j = I_{rand} \\ X_{j,i,G+1} & \text{if } rand_{j,i} > CR \quad or \quad j \neq I_{rand} \end{cases} \tag{16}$$

With $rand_{j,i} \sim U(0,1)$, I_{rand} is a random integer from $(1, 2, ..., D)$ where D is the solution's dimension i.e. number of control variables. I_{rand} ensures that $V_{i,G+1} \neq X_{i,G}$.

Selection operation

To keep the population size constant over subsequent generations, selection operation is performed. In this operation the target vector $X_{i,G}$ is compared with the trial vector $V_{i,G+1}$ and the one with the better fitness value is admitted to the next generation. The selection operation in DE can be represented by:

$$X_{i,G+1} = \begin{cases} U_{i,G+1} & \text{if } f(U_{i,G+1}) < f(X_{i,G}) \\ X_{i,G} & \text{otherwise.} \end{cases} \tag{17}$$

where $i \in [1, NP]$.

RESULTS AND DISCUSSIONS

Implementation of DE

The effectiveness, efficiency, and robustness of the DE algorithm are sensitive to the settings of the control parameters. The control parameters in DE are step size function also called scaling factor (F), crossover probability (CR), the number of population (N_P), initialization, termination and evaluation function. F controls the amount of perturbation in the mutation process and generally lies in the range $(0, 1)$. Crossover probability (CR) constants are generally chosen from the interval $(0.5, 1)$. Several strategies can be employed in DE optimization algorithm. The strategy in a DE algorithm is denoted by $DE/x/y/z$, where x represents the mutant vectors, y represents the number of difference vectors used in the mutation process and z represents the crossover scheme used in the crossover operation. The suggested choice of control parameters is [25] population size of $N_P = 50$ ($N_P = 5D$ where D = dimensionality of the problem), step size $F = 0.8$ and crossover probability of $CR = 0.8$ and these values are selected in the present paper. The strategy employed is as follows: DE/best/1/exp. Optimization is terminated by the pre-specified number of generations which is set to 100. The flow chart of the DE algorithm employed in the present study is given in Fig. 6. The model of the system under study shown in Fig. 1 is developed in MATLAB/SIMULINK environment and DE program is written (in .mfile). Initially, fuzzy PID controllers without TCSC and SMES units are considered for each area. Scaling factors and PID controller gains are chosen in the range [0 −2] and [−2 2] respectively. The developed model is simulated in a separate program (by .mfile using initial population/controller parameters) considering a 1% step load change in area 1. The objective function (ITAE) value for each individual is calculated in the SIMULINK model file and transferred to .mfile through workspace. These objective function values are used to assess the populations. The population is then modified by applying mutation, crossover and selection operators in the main DE program as given in Flow chart (Fig. 6). Simulations were conducted on an Intel, core i-3core cpu, of 2.4 GHz and 4 GB RAM computer in the MATLAB 7.10.0.499 (R2010a) environment. The optimization was repeated 50 times and the best final solution among the 50 runs is chosen as proposed controller parameters. The best final solutions obtained in the 50 runs are shown in Table 2.

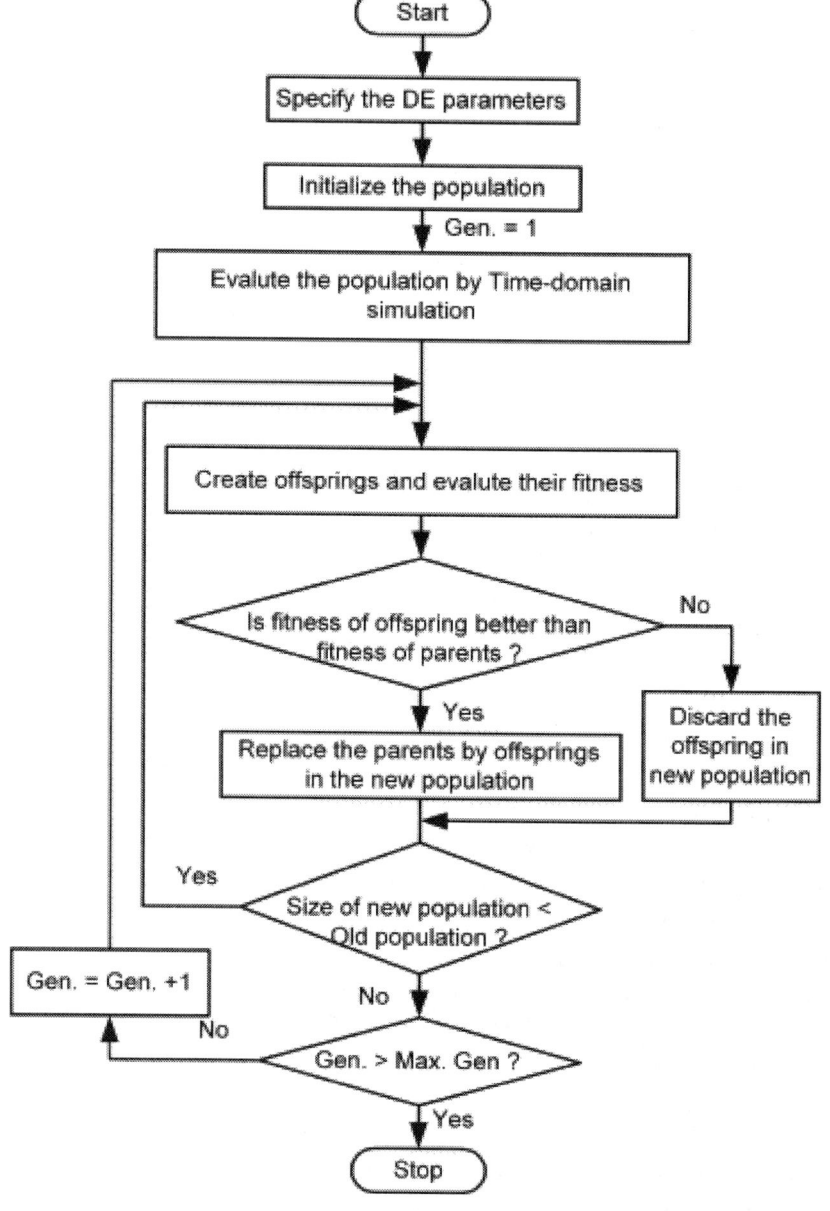

Figure 6. Flow chart of proposed DE optimization approach.

Table 2. Tuned fuzzy PID controller parameters.

Optimum controller gains	Genetic Algorithm (GA)	Differential Evolution (DE)		
	Without TCSC and SMES	Without TCSC and SMES	With TCSC	With TCSC and SMES
K_1	1.5553	1.8589	0.1943	0.1132
K_2	1.792	1.9092	1.5638	1.682
K_{P1}	−0.0805	0.5481	1.672	1.5621
K_{I1}	1.346	1.9461	1.7881	1.5075
K_{D1}	1.3483	0.517	−0.6707	−0.3560
K_3	1.9128	0.3338	0.5414	0.1932
K_4	0.2626	0.3607	0.2572	0.3662
K_{P2}	0.2813	− 0.2837	−0.9896	−1.7376
K_{I2}	1.1323	−0.8919	−0.2052	1.3894
K_{D2}	−0.0970	1.3173	0.9608	−1.0557

Analysis of results

The objective function (ITAE) value given by Eq. (3) is determined by simulating the developed model by applying a 1% step increase in load in area 1. The corresponding performance index in terms of ITAE value, settling times (2%) and peak overshoots in frequency and tie line power deviations is shown in Table 3. For comparison, the corresponding values of GA optimized fuzzy PID controllers are also shown in Table 3. For the implementation of GA, normal geometric selection, arithmetic crossover and non-uniform mutation are employed in the present study. A population size of 50 and maximum generation of 100 is employed in the present paper. A detailed description about GA parameters employed in the present paper can be found in reference [9]. It should be noted here that, GA values correspond to same power system, controller structure (fuzzy PID) and objective function employed (ITAE) for proper comparison of techniques. It is evident from Table 3 that DE outperform GA as minimum ITAE value is obtained with DE (ITAE = 1.2250) compared to GA (ITAE = 2.1429). The dynamic performance of the system is shown in Figure 7, Figure 8 and Figure 9 for 1% step increase in load in area 1. It is clear from Figure 7, Figure 8 and Figure 9 that better dynamic performance is obtained by DE optimized fuzzy PID controller compared to GA optimized fuzzy PID controller. Hence it can be concluded that DE outperform GA technique.

Table 3. Comparative performance of error and settling time.

Parameters	ITAE	Settling time (2% band), T_s (s)			Peak over shoot ($\times 10^{-3}$)		
		ΔF_1	ΔF_2	ΔP_{Tie}	ΔF_1	ΔF_2	ΔP_{Tie}
GA	2.1429	37.05	34.85	34.06	6.413	1.568	0.839
DE	1.225	33.26	35.03	30.43	3.974	1.528	0.756
DE: with TCSC	0.8178	21.84	22.25	26.25	3.768	8.01	0.762
DE: both TCSC and SMES	**0.6672**	**21.13**	**21.57**	**16.85**	**2.5113**	**5.44**	**0.29**

The bold values are indicates the best results.

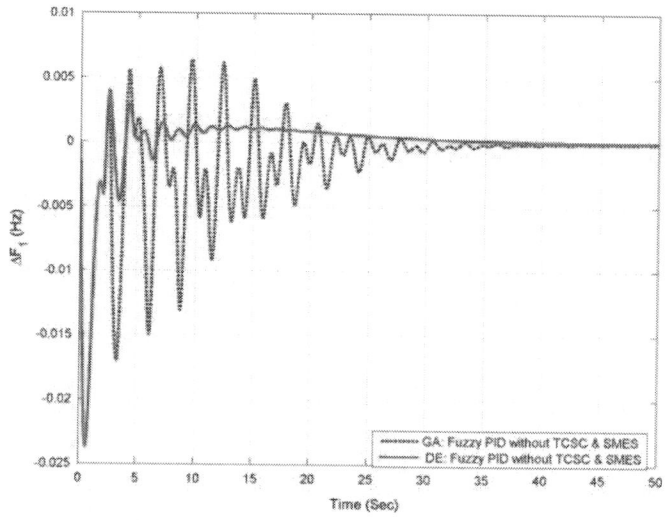

Figure 7. Change in frequency of area-1 for 1% load change in area-1 without TCSC and SMES units.

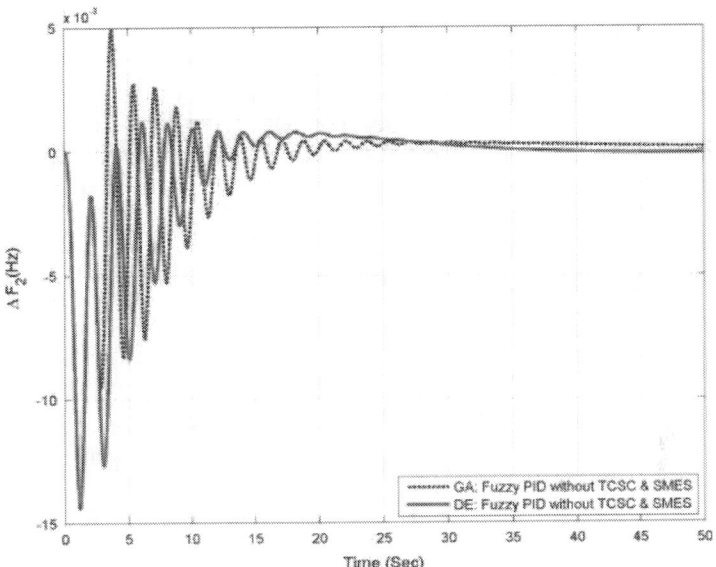

Figure 8. Change in frequency of area-2 for 1% load change in area-1 without TCSC and SMES units.

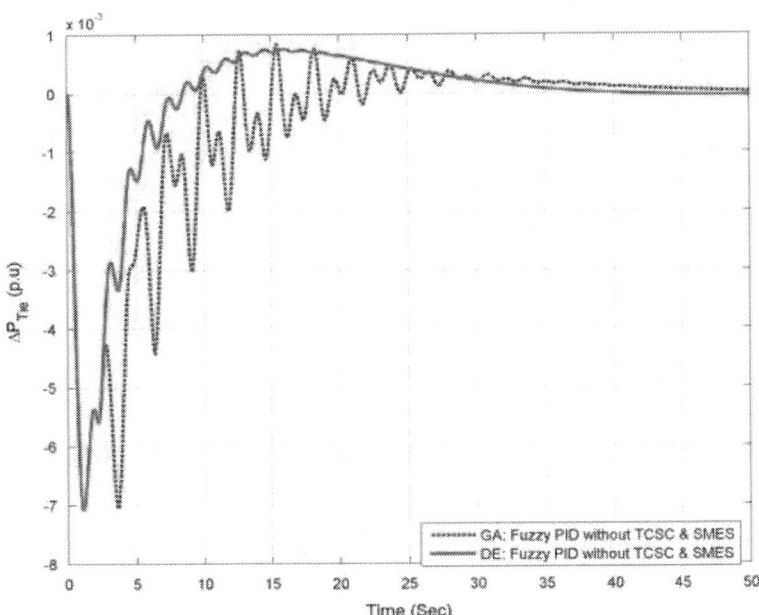

Figure 9. Change in tie line power for 1% load change in area-1 without TCSC and SMES units.

In the next step, the TCSC is incorporated separately in the tie-line to analysis its effect on the power system performance. Subsequently SMES units are installed in both areas and coordinated with TCSC to study their effect on system performance. The results of fuzzy PID controller with TCSC employing differential evolution algorithm over 50 independent runs are shown in Table 2. It is clear from Table 3 that by employing the TCSC along with fuzzy PID controller, the objective function (ITAE) value is decreased to 0.8178 (i.e. 33.24% improvement). In addition better results are observed in terms of settling time and peak overshoot values with the TCSC fuzzy PID compared to without TCSC. It is also seen that with coordinated application of TCSC and SMES units, the ITAE value is further reduced to **0.6672**. It can be seen from Table 3 that with TCSC and SMES, the settling times of ΔF_1, ΔF_2 and ΔP_{Tie} are improved compared to others for the same investigated system with similar objective function (ITAE).

To study the dynamic performance of the system a step increase in demand of 1% is applied at $t = 0$ s in area-1 and the system dynamic responses are shown in Figure 10,Figure 11 and Figure 12. Critical analysis of the dynamic responses clearly reveals that significant system performance improvement in terms of minimum undershoot and overshoot in frequency oscillations as well as tie-line power exchange is observed with coordinated application of TCSC and SMES units.

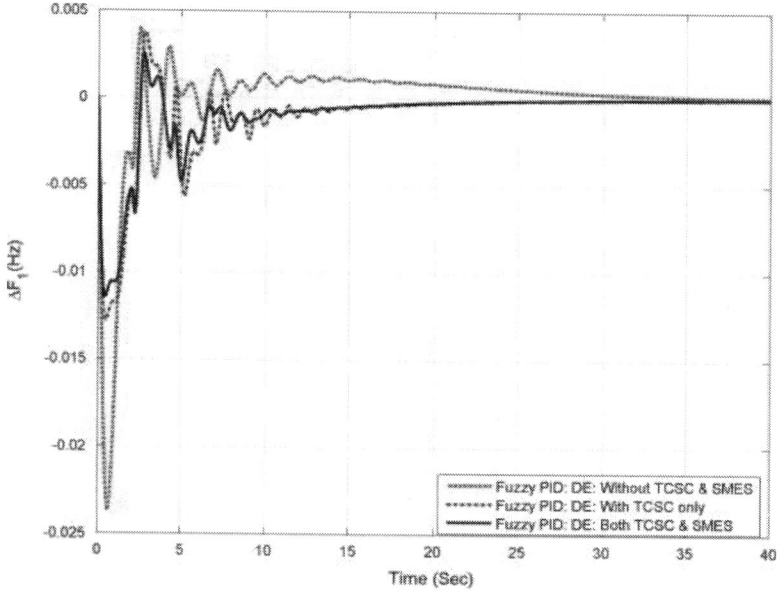

Figure 10. Change in frequency of area-1 for 1% load change in area-1.

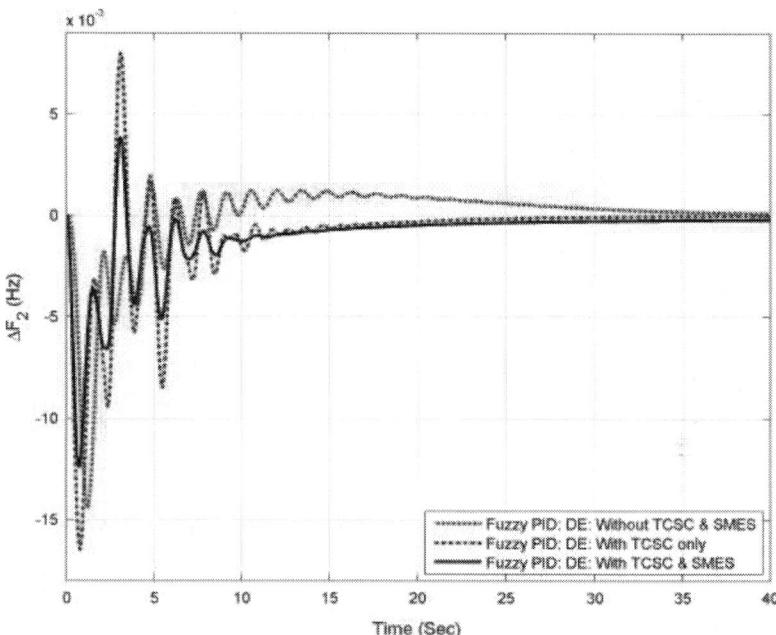

Figure 11. Change in frequency of area-2 for 1% load change in area-1.

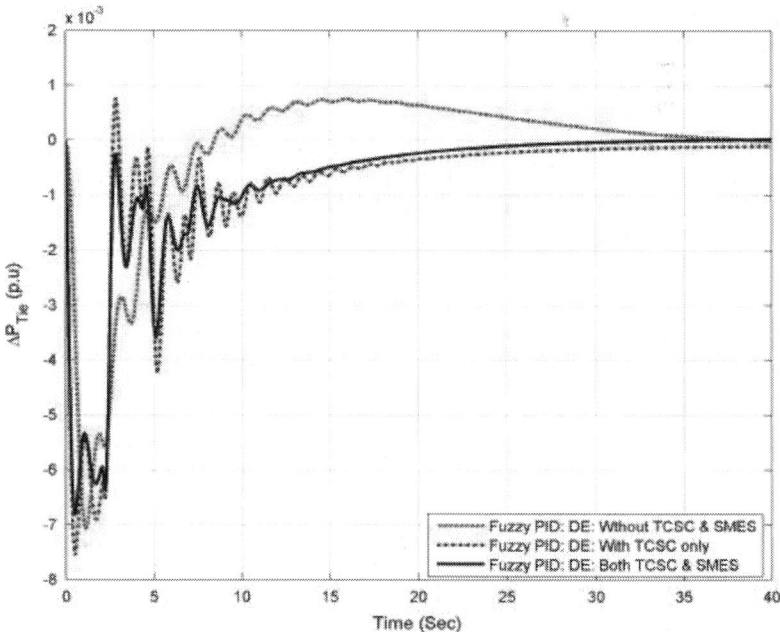

Figure 12. Tie-line power deviation for 1% load change in area-1.

Sensitivity analysis

Sensitivity analysis is carried out to study the robustness the system to wide changes in the operating conditions and system parameters [5], [27] and [28]. In this section robustness of the power system is checked by varying the loading conditions and system parameters from their nominal values (given in Appendix A) in the range of +25% to −25% without changing the optimum values of fuzzy PID controller gains. The change in operating load condition affects the power system parameters K_P and T_P. The power system parameters are calculated for different loading conditions as given in Appendix A. The system with TCSC and SMES units are considered in all the cases due to their superior performance. The various performance indexes (settling time, peak overshoot and ITAE) under normal and parameter variation cases for the system are given in Table 4. It can be observed from Table 4 that settling time, peak overshoot and ITAE values vary within acceptable ranges and are nearby equal to the respective values obtained with nominal system parameter. It is also evident from Table 5 and Table 6 that the eigenvalues lie in the left half of s-plane for all the cases thus maintain the stability. Hence, it can be concluded that the proposed controllers are robust and perform satisfactorily when system parameters changes in the range ±25%. The dynamic performance of the system with the varied conditions of loading, T_G, T_T, B and R is shown in Figure 13, Figure 14, Figure 15, Figure 16, Figure 17, Figure 18 and Figure 19. It can be observed from Figure 13, Figure 14, Figure 15, Figure 16, Figure 17, Figure 18 and Figure 19 that the effect of the variation in loading condition and system parameters on the system performance is negligible. Hence the optimum values of controller parameters obtained at the nominal loading with nominal parameters, need not be reset for wide changes in the system loading or system parameters.

Table 4. Sensitivity analysis.

Parameter variation	% Change	Performance index with TCSC and SMES						ITAE
		Settling time, T_s (s)			Peak over shoot × 1			
		ΔF_1	ΔF_2	ΔP_{Tie}	ΔF_1	ΔF_2	ΔP_{Tie}	
Nominal	0	**21.1**	**21.6**	**16.85**	**2.511**	**5.44**	**0.29**	**0.667**
Loading condition	25	21.8	22.2	17.15	2.489	5.36	0.266	0.665
	−25	20.5	20.9	16.4	2.533	5.52	0.313	0.683
T_G	25	23	23.5	17.59	2.835	5.9	0.488	0.714
	−25	19.3	19.8	15.87	2.147	4.95	0.171	0.735
T_t	25	19.4	19.9	15.94	2.564	5.48	0.298	0.72
	−25	23.3	23.7	17.68	2.395	5.29	0.233	0.726
B	25	20.6	21.1	17.37	2.493	5.53	0.299	0.649
	−25	21.7	22	16.14	2.531	5.35	0.28	0.691
R	25	21.5	21.9	17.11	2.494	5.49	0.194	0.707
	−25	20.7	21.1	16.32	2.514	5.36	0.422	0.61

The bold values are indicates the best results.

Table 5. System eigen values under parameter variation in loading, T_G and T_T with TCSC and SMES units.

Loading condition		T_G		T_T	
25%	−25%	25%	−25%	25%	−25%
−48.2086	−48.2096	−48.2093	−48.2086	−48.2093	−48.2088
−31.9578	−31.9592	−31.9596	−31.9561	−31.9593	−31.9572
−33.0234	−33.0237	−33.0236	−33.0236	−33.0236	−33.0236
−13.4364	−13.4354	−13.4629	−16.9560	−13.4497	−13.4189 ± 0.0790i
−13.2648	−13.2628	−12.9373	−13.4694	−13.1737	−1.4553 ± 4.7764i
−1.3825 ± 4.7594i	−1.3708 ± 4.7614i	−10.3946	−12.8478	−1.3341 ± 4.7430i	−1.2922 ± 3.0029i
−1.3825 ± 2.8736i	−1.2456 ± 2.8684i	−1.3570 ± 4.7719i	−1.3943 ± 4.7438i	−1.2419 ± 2.7746i	−2.7612
−1.3360	−1.3376	−1.2261 ± 2.8569i	−1.2809 ± 2.8822i	−2.1386	−1.3774
−0.1269	−0.1271	−1.3324	−1.3411	−1.2845	−0.1269
−0.0107	−0.0107	−0.1270	−0.1270	−0.1271	−0.0107
−0.0047	−0.0047	−0.0107	−0.0107	−0.0107	−0.0047
−0.1000	−0.1000	−0.0047	−0.0047	−0.0047	−2.3810
−12.4999	−12.5000	−12.5000	−12.5000	−2.3810	−0.1000
−12.5000	−12.5000	−0.1000	−0.1000	−0.1000	−12.5000
−2.3809	−2.3810	−2.3810	−2.3810	−12.5000	−12.5000
−2.3809	−2.3810	−2.3810	−2.3810	−12.5000	

Table 6. System eigen values under parameter variation in B and R with TCSC and SMES units.

B		R	
25%	−25%	25%	−25%
−48.2091	−48.2091	−48.2093	−48.2088
−31.9620	−31.9556	−31.9592	−31.9574
−32.9426	−33.1037	−33.0236	−33.0236
−13.4426	−13.4293	−13.4482	−13.4023 ± 0.0620i
−13.2631	−13.2646	−13.1869	−1.4277 ± 4.8201i
−1.3964 ± 4.7793i	−1.3573 ± 4.7410i	−1.3443 ± 4.7293i	−1.1697 ± 2.9829i
−1.2685 ± 2.8656i	−1.2363 ± 2.8767i	−1.3022 ± 2.7924i	−1.2936
−1.3362	−1.3374	−1.3679	−0.1305
−0.1270	−0.1270	−0.1247	−0.0109
−0.0051	−0.0104	−0.0106	−0.0043
−0.0111	−0.0043	−0.0050	−0.1000
−0.1000	−0.1000	−0.1000	−12.5000
−12.5000	−12.5000	−12.5000	−12.5000
−12.5000	−12.5000	−12.5000	−2.3810
−2.3810	−2.3810	−2.3810	−2.3810
−2.3810	−2.3810	−2.3810	

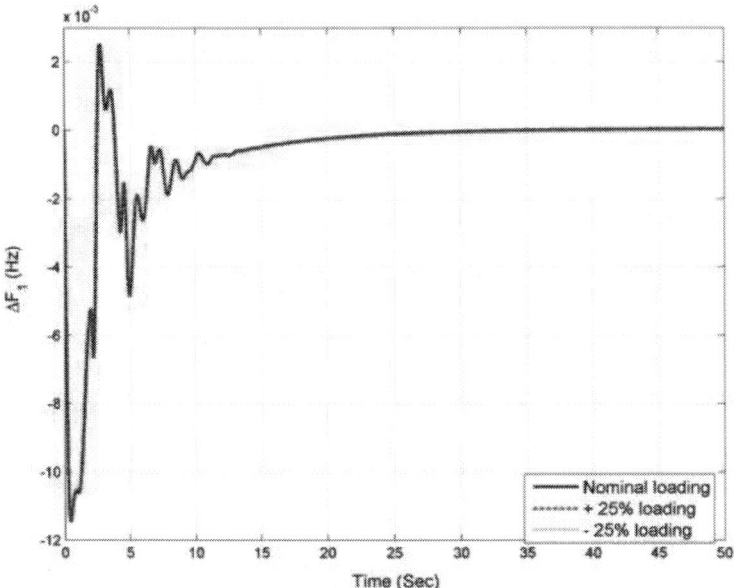

Figure 13. Change in frequency of area-1 for 1% load change in area-1 with variation in loading.

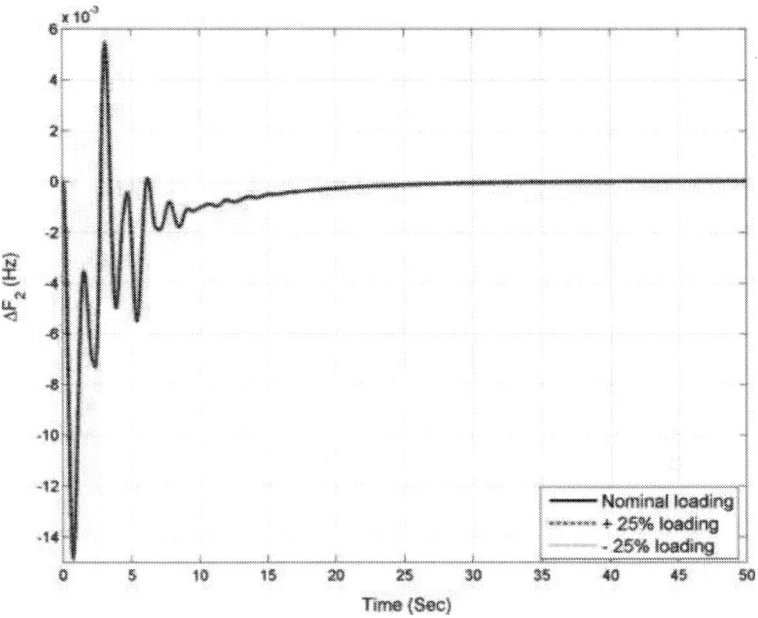

Figure 14. Change in frequency of area-2 for 1% load change in area-1 with variation in loading.

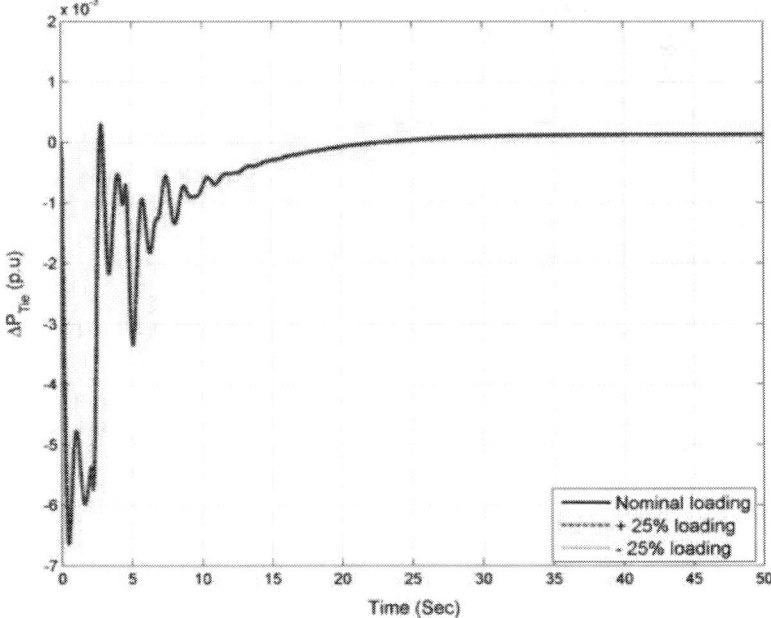

Figure 15. Tie-line power deviation for 1% load change in area-1 with variation in loading.

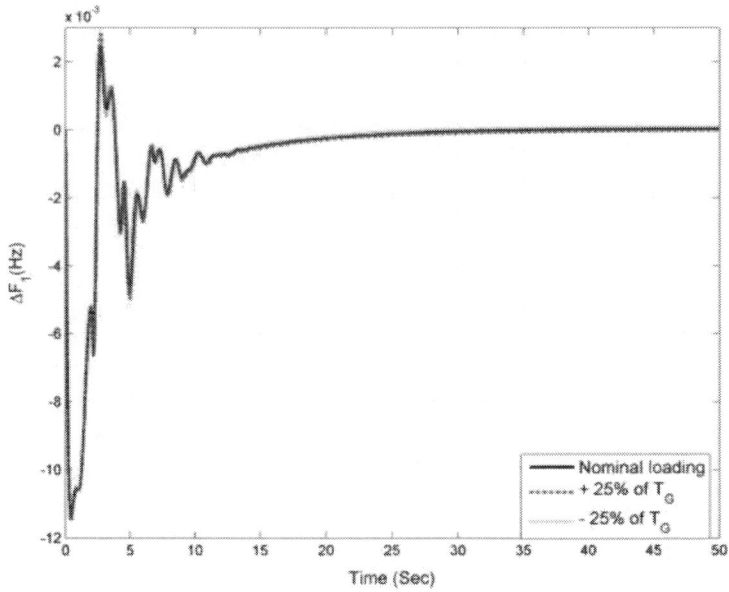

Figure 16. Change in frequency of area-1 for 1% load change in area-1 with variation in T_G.

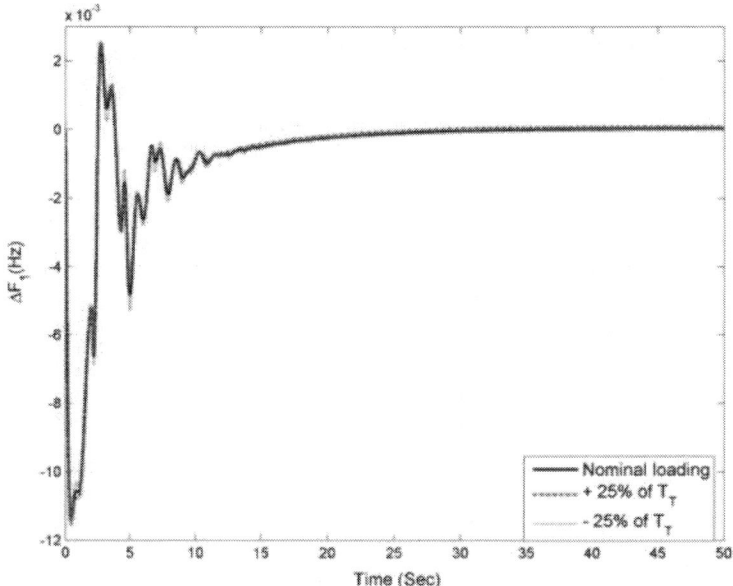

Figure 17. Change in frequency of area-1 for 1% load change in area-1 with variation in T_T.

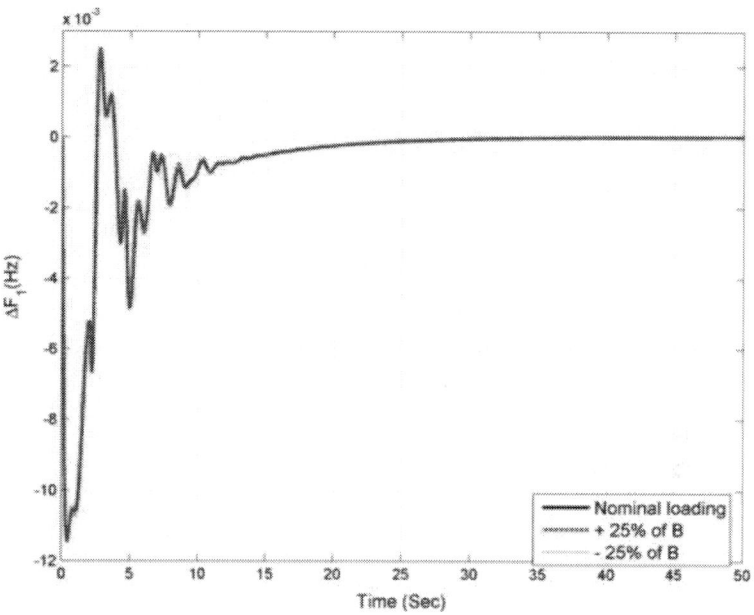

Figure 18. Change in frequency of area-1 for 1% load change in area-1 with variation in B.

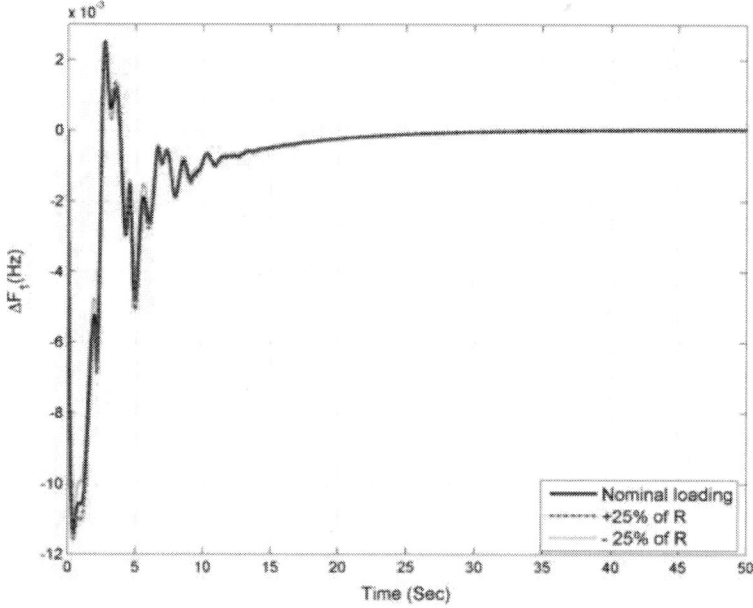

Figure 19. Change in frequency of area-1 for 1% load change in area-1 with variation in R.

CONCLUSION

In this paper, a Differential Evolution (DE) algorithm optimized fuzzy PID controller has been proposed for Automatic Generation Control (AGC) of multi-area multi-units power systems. Initially a multi-area multi-units power system with the considerations of physical constraints such as GRC and time delays is considered and the superiority of DE over GA is demonstrated. A linear incremental model for a TCSC has also been developed which is suitable for AGC applications. Further, TCSC and SMES units are added in the system model in order to improve the system performance. It is observed that when the TCSC unit is placed with the tie-line, dynamic performance of system is improved. Then the impact of SMES units in the AGC along with TCSC is studied. From the simulation results, it is observed that significant improvements of dynamic responses are obtained with coordinated application of TCSC and SMES units. Finally, sensitivity analysis is carried out to show the robustness of the controller by varying the loading conditions and system parameters in the range of +25% to −25% from their nominal values. For systems under study, it is revealed that the parameters of the proposed DE optimized fuzzy PID controllers need not be reset even if the system is subjected to wide variation in loading conditions and system parameters.

APPENDIX A

Nominal parameters of the system investigated are:

(i) **Multi-area multi-units system**

$$B_1, B_2 = 0.42249 \text{ p.u. MW/Hz}; \ R_1 = R_2 = R_3 = R_4 = 2.4 \text{ Hz/p.u.}; \ T_{G1}$$
$$= T_{G2} = T_{G3} = T_{G4} = 0.08 \text{ s}; \ T_{T1} = T_{T2} = T_{T3} = T_{T4} = 0.3 \text{ s}; \ K_P$$

(ii) $\quad = 120 \text{ Hz/p.u.}; \ TP = 20 \text{ s}; K_{R1} = K_{R2} = 10; \ T_{R1} = T_{R2} = 10 \text{ s}$

(iii) **TCSC data**

$$T_{12} = 0.0866; \ \delta_0 = 300; \ X_t = 10 \text{ p.u.}; \ K_{TCSC} = 2.0;$$

(iv) $T_{TCSC} = 0.02 \text{ s}$

(v) **SMES data**

(vi) $K_{SMES} = 0.12;\ T_{SMES} = 0.03$ s

REFERENCES

1. Kundur P. Power system stability and control. New York: McGraw-Hill; 1994.
2. Elgerd OI. Electric energy systems theory – an introduction. 2nd ed. Tata McGraw Hill; 2007.
3. Kothari DP, Nagrath IJ. Modern power system analysis. 4th ed. New Delhi: Tata McGraw-Hill; 2011.
4. Saikia LC, Nanda J, Mishra S. Performance comparison of several classical controllers in AGC for multi-area interconnected thermal system. Int J Electr Power Energy Syst 2011;33:394–401.
5. Parmar KPS, Majhi S, Kothari DP. Load frequency control of a realistic power system with multi-source power generation. Int J Electr Power Energy Syst 2012;42:426–33.
6. Saikia LC, Mishra S, Sinha N, Nanda J. Automatic generation control of a multi area ydrothermal system using reinforced learning neural network controller. Int J Electr Power Energy Syst 2011;33(4):1101–8.
7. Ibraheem KP, Kothari DP. Recent philosophies of automatic generation control strategies in power systems. IEEE Trans Power Syst 2005;20(1):346–57.
8. Ghosal SP. Optimization of PID gains by particle swarm optimization in fuzzy based automatic generation control. Electr Power Syst Res 2004;72(3):203–12.
9. Golpira H, Bevrani H, Golpira H. Application of GA optimization for automatic generation control design in an interconnected power system. Energy Convers Manage 2011;52:2247–55.
10. Yesil E, Guzelkaya M, Eksin I. Self tuning fuzzy PID type load and frequency controller. Energy Convers Manage 2004;45(3):377–90.
11. Khuntia SR, Panda S. Simulation study for automatic generation control of a multi-area power system by ANFIS approach. Appl Soft Comput 2012;12(1):333–41.

12. Hingorani NG, Gyugyi L. Understanding FACTS-concepts and technology of flexible AC transmission systems. Standard Publishers, IEEE Press; 2000.

13. Praghnesh B, Ghoshal SP, Ranjit R. Load frequency stabilization by coordinated control of thyristor controlled phase shifters and superconducting magnetic energy storage for three types of interconnected two-area power systems. Int J Electr Power Energy Syst 2010;32:1111–24.

14. Mathur RM, Varma RK. Thyristor-based FACTS controllers for electrical transmission systems. IEEE Press, John Wiley & Sons, inc. publication; 2002.

15. Stron R, Price K. Differential evolution – a simple and efficient adaptive scheme for global optimization over continuous spaces. J Global Optim 1995;11:341–59.

16. Das S, Suganthan PN. Differential evolution: a survey of the state-of-the-art. IEEE Trans Evol Comput 2011;15:4–31.

17. Brest J, Greiner S, Boskovic B, Mernik M, Zumer V. Selfadapting control parameters in differential evolution: a comparative study on numerical benchmark problems. IEEE Trans Evol Comput 2005;10:646–57.

18. Cheres E. The application of generation rate constraint in modeling of a thermal power system. Electr Power Comp Syst 2001;29(2):83–7.

19. Ignacio E, Fernandez-Bernal F, Rouco L, Elosia P, Saiz-Chicharro A. Modeling of thermal generating units for automatic generation control purposes. IEEE Trans Control Syst Technol 2004;12(1):205–10.

20. Panda S. Differential evolution algorithm for SSSC-based damping controller design considering time delay. J. Franklin Inst 2011;348(8):1903–26.

21. Mudi KR, Pal RN. A robust self-tuning scheme for PI-and PDtype fuzzy controllers. IEEE Trans Fuzzy Syst 1999;7(1):2–16.

22. Woo WZ, Chung YH, Lin JJ. A PID type fuzzy controller with self tuning scaling factors. Fuzzy Sets Syst 2000;115(2):321–6.

23. Shabani H, Vahidi B, Ebrahimpour M. A robust PID controller based on imperialist competitive algorithm for load–frequency control of power systems. ISA Trans. 2012;52:88–95.

24. Sudha KR, Vijaya SR. Load frequency control of an interconnected reheat thermal system using type-2 fuzzy system including SMES units. Int J Electr Power Energy Syst 2012;43:1383–92.

25. Janez B, Saso G, Borko B, Marjan M, Viljem Z. Self-adapting control parameters in differential evolution: a comparative study on numerical benchmark problems. IEEE Trans Evol Comput 2005;10:646–57.
26. Qin AK, Huang VL, Suganthan PN. Differential evolution algorithm with strategy adaptation for global numerical optimization. IEEE Trans Evol Comput 2009;13:398–417.
27. Sahu RK, Panda S, Rout UK. DE optimized parallel 2-DOF PID controller for load frequency control of power system with governor dead-band nonlinearity. Int J Electr Power Energy Syst 2013;49:19–33.
28. Rout UK, Sahu RK, Panda S. Design and analysis of differential evolution algorithm based automatic generation control for interconnected power system. Ain Shams Eng J 2013;4(3):409–21.

CITATION

Saroj Padhan, Rabindra Kumar Sahu, Sidhartha Panda, Automatic generation control with thyristor controlled series compensator including superconducting magnetic energy storage units, Ain Shams Engineering Journal, Volume 5, Issue 3, September 2014, Pages 759-774, ISSN 2090-4479, http://dx.doi.org/10.1016/j.asej.2014.03.011.

CHAPTER 7

Fuzzy Controller for Automatic Steering in Heavy Vehicle Semi-Trailers

L. Bortoni-Anzures[1], D. Gómez-Meléndez[1], G. Herrera-Ruíz[1], M. Martínez-Madrid[2]

[1]Universidad Autónoma de Querétaro, Facultad de Ingeniería
[2]Instituto Mexicano del Transporte Tamaulipas, México

ABSTRACT

Trucks with semi-trailers are widely used for transportation of goods due their low operation cost, but inherent to these vehicles are some problems such as a poor maneuverability. To minimize the effects of this disadvantage, among other solutions, the incorporation of steerable axles in the semitrailers has been proposed. This paper presents a steering equation, and a fuzzy-logic controller for a semi-trailer automatic forced-steering system to minimize the off-tracking and the total swept path width, resulting in an improvement of vehicle's maneuverability at low speeds. To accomplish this, the suggested control algorithm considers the articulation angle and parameters such as vehicle speed and direction. The system was tested on an instrumented experimental semi-trailer during various predetermined test maneuvers.

INTRODUCTION

Articulated vehicles have proven their economic profitability, but as the number of these vehicles grows, it becomes evident that there is a substantial need to improve their handling control performance.

Different approaches have been taken into consideration; one important and promising is related to trajectory matching and lateral control of these heavy vehicles. In this line of work the objective is, in general, to make the articulated vehicles able to follow a secure and efficient path (Bolzern and Locatelli, 2001; El-Gindy, 1978; Hingwe *et al.*, 2000;Sampei *et al*, 1995 and Tsao *et al.*, 2006), are a few of researchers who specialize in trajectory matching.

The biggest project in this area is the California Partners for Advanced Transit and Highways (California Paths) (Hingwe *et al.*, 2000) where numerous groups of government, public and private institutions are involved in the development of automated highway systems (AHS).

A different approach is to propose design modifications or the implementation of mechanical components that improve overall vehicular maneuverability performance. Steerable axles on semi-trailers are an option to, not only improve maneuverability, but also reduce risk of accidents and decrease fuel consumption and tire wear. Further, it may also minimize hazards and damage to roadway infrastructures. Several mechanical configurations have been tested via computer simulation (Sankar *et al.*, 1991), yet few of them are already used as prototypes or are in early commercial stages.

Those mechanisms represent significant benefits for vehicles, but they still need to be improved since some stability problems have been raised. Inside the group of steering axles of a semitrailer, the command-steer system is the most efficient option (Jujnovich and Cebon, 2002).

In 2000 an Australian company, Gayat Pty. Ltd. (Prem and Ramsay, 2001) and in 2005 the Cambridge Vehicle Dynamics Consortium

(Cambridge University) (Neads, 2006) presented full scale prototypes of semi-trailers with command steer axles systems, each of them is part of the evolution of this technology.

Moreover, results from the tests carried out for this paper present work on an actual experimental semi-trailer. These results also show a control equation with a proposed method of employment for a system of control based on fuzzy-logic that could further advance in command steering technology.

MATERIALS AND METHODS

An experimental semi-trailer was made in order to obtain a better understanding of the articulated vehicles maneuverability performance and the most significant characteristics affecting it.

Pulled by a pickup truck, this semi-trailer which is shown in Figure 1 incorporates several configurable characteristics such as the location on the fifth wheel (G) in front, on top or in the back from the traction axle. The length of the semi-trailer (H) and an axle in the semi-trailer that is configurable as fixed or steerable, allowing the testing of different control schemes.

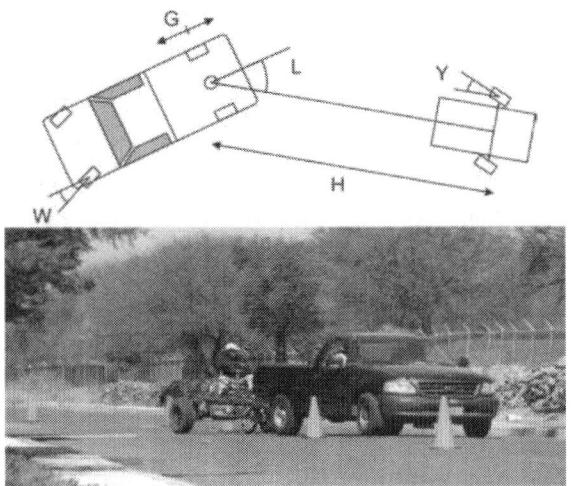

Figure 1. Experimental vehicle

Instrumentation

To track the geometric behavior of the vehicle, a GPS receiver was installed on top of the fifth wheel, and angular transducers in the steering wheels of the truck, in the hitch and in the steering wheels of the semi-trailer. This procedure is clearly explained in (Bortoni *et al.*, 2007).

Also accurate yaw rate sensors and lateral accelerometers were placed in the chassis of the truck and the semi-trailer. Finally a mini camera was installed in the front of the truck pointed toward the ground to guide the driver during specific maneuvers (Figure 2).

Figure 2. Video camera and video camera view at marks at the front of the truck

The signals were monitored by two portable computers. The sample rate was established in 50 samples per second per channel. This sampling frequency was considered high enough to have a detailed time history of the measurements.

Handling performance

Handling performance is the result of a vehicle's steering inputs determined by the vehicle weights, dimensions, and mechanical properties of the tires, suspension, vehicle frames, and weight distribution among the axles (Sampson, 2000).

Consequently, a number of performance measures can be compiled to assess the maneuverability performance of the vehicles, particularly with regard to articulated trucks. This study considers the performance measures related to geometrical requirements, as:

Off-tracking

Whenever a vehicle with more than one axle turns, rear wheels fail to follow the path of front wheels precisely. Off tracking measures the maximum distance between the paths of the steering axle and the axle of the most rearward wheels (Battelle Team, 1995).

Swept path width

Jujnovich and Cebon (2002) describe the "swept path width" as the maximum width of the swept trajectory in a small radius turn. Figure 3 shows both, the off-tracking and swept path width measurements during circular constant radius turns, also called stationary turns.

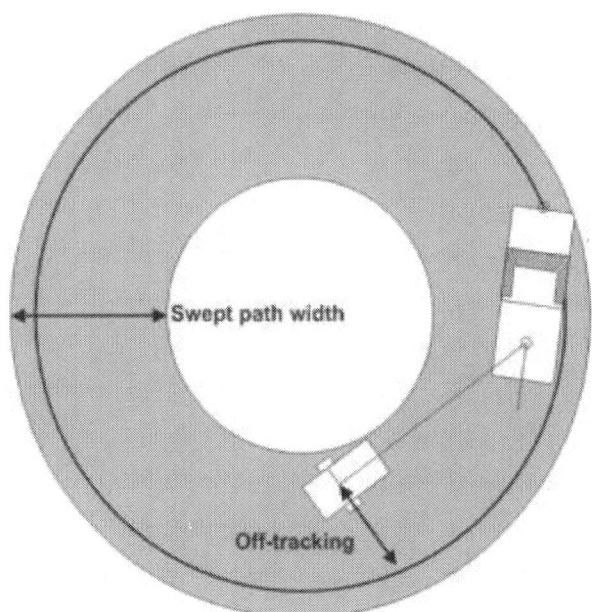

Figure 3. Off-tracking and swept path width during a stationary turn

Test maneuvers

For testing, three different maneuvers were considered: two related to a predetermined defined following path and one slalom type. The first two consisted of a stationary turn and a 90° transient turn, performed at a very low velocity.

Circular and transient curve paths were traced on the pavement surfaces which were to be followed by the truck to facilitate the path following an 8, 9.5 and 11.25 m. radius, the stationary turn like the illustration in Figure 3 and transient turn in Figures 4 and 5.

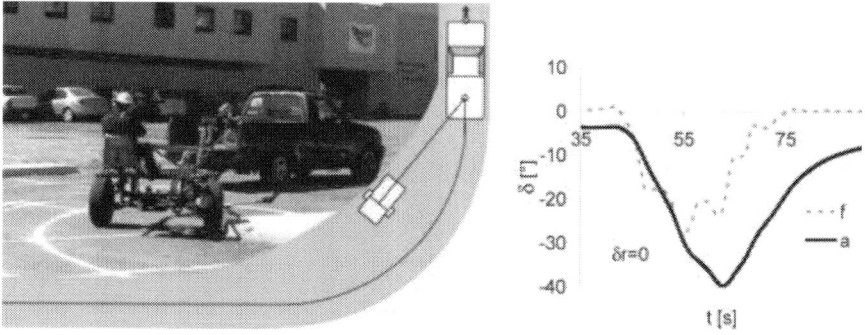

Figure 4. Transient turn with fixed semi-trailer axle

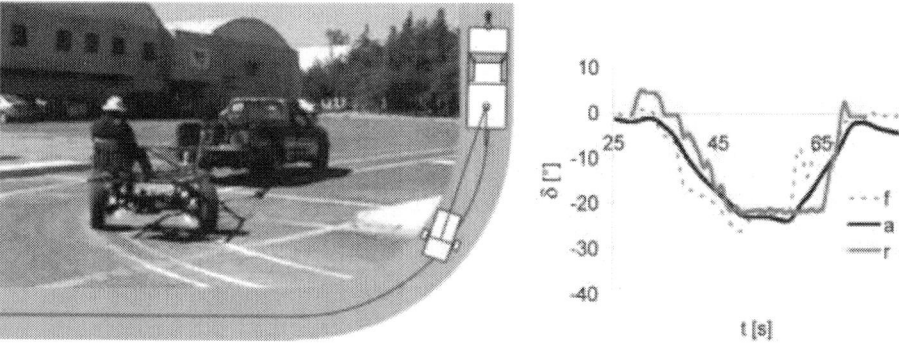

Figure 5. Transient turn with command steering semi-trailer axle

Figure 4 shows the typical maneuverability performance of an articulated vehicle. f stands for the angle of the front axle, the steering axle on the truck, and **a** for the articulation angle.

In the case when the 9.5 m maneuver was made, the results observed demonstrated that the angle of the articulation reached 40° which means an off-tracking of 2.95 m and a swept path width of 4.6 m.

When the semi-trailer axle is set to command steer as seen in Figure 5, it was found that the maneuver results in a maximum articulation angle of barely 22°, which is an improvement of approximately 50%, therefore resulting in a negative off-tracking and a swept path width that equals that of the width of the vehicle.

The line **r** represents the angle of the semi-trailer axle, and as Figure 5 shows, the driver in the semi-trailer performs a momentary opposite turn to get better alignment with the truck. When it leaves the transient part of the maneuver, in this case a 20° turn, the semitrailer axle was applied.

Besides the different radius, these maneuvers were performed from the left and from the right, under different vehicular configurations as the longitude of the semi-trailer and the fifth wheel location.

The slalom type maneuver consists of a series of directional changes around vial cones that were placed 10 meters apart on a street that was 7 m wide while maintaining a constant speed of 8 km/h. Figure 1 shows the vehicle during the slalom maneuver.

In these maneuvers there is not a defined trajectory, so the driver relies on his perception and expertise to deal with the obstacles. In this case the way to evaluate the maneuver is directly related to the lateral speed of the truck in contrast with the one on the semi-trailer.

In Figure 6A, the lateral speed of the truck is almost twice that of the semi-trailer. This means that the truck is under high demand for the vehicle to avoid obstacles. On the other hand, in Figure 6B when the command steering was activated the speed of the truck and the semi-trailer, is almost the same, requiring less lateral space.

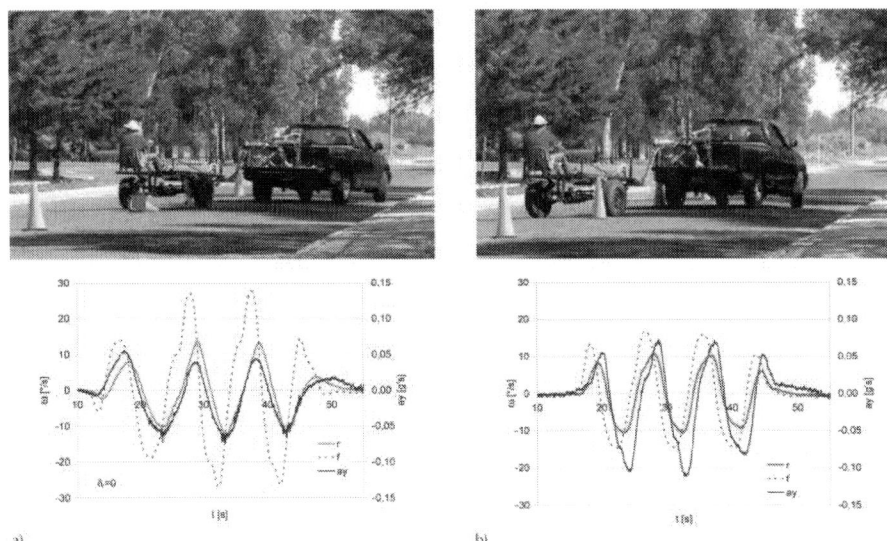

Figure 6. Shows yaw rates and semi-trailer's axle lateral acceleration history, a) is the fixed semi-trailer axle, while b) is the steerable semi-trailer's axle

The difference in magnitude in the left verses the right turn corresponds to the perception of the driver. Note: It is easier to follow the obstacle and the edge on the driver's side. Also, in the pictures can be noticed de difference in road space demand, with the fixed axle semi-trailer, the vehicle use the full wide of the road, and even then it fails to deal with all of the obstacles. While the command steer semi-trailer succeed all obstacles occupying less space.

There is a significant amount of literature regarding studies comparing the behavior between fixed axles and steering command ones using simulation models. In most cases it was concluded that by incorporating steering axles important benefits can be obtained such as improving handling capabilities of the vehicle and reduction of tire friction demand (Jujnovich and Cebon, 2002; Billing and Patten, 2003).

Test results using an experimental vehicle confirm this, but it also extends the scope of testing beyond most simulation results and illustrates some aspects of the driver's perception on the experimental assessments which in turn affects the maneuver performance.

Control Equation

During previous tests on full scale vehicles (Figure 7) and from the Ackerman equations, the angle in the semi-trailer axle that would minimize the off-tracking for each radios turn were geometrically calculated This defines the general behavior required for the specific semi-trailer dimensional configuration.

Figure 7. Test vehicles used in the definition of the control equation

The same dimensional characteristics were considered in the construction of the experimental semi-trailer, in order to fine tune the control equation and to later experiment with the width the parameters in the algorithm of control.

After this, by combining the dimensional characteristics of the semi-trailer with the Ackerman equations, an evaluation computer program was set up to track the off-tracking results during the test maneuvers period and by using interpolation regression techniques, a control equation was defined (Eq. 1 and 2, Figure 8).

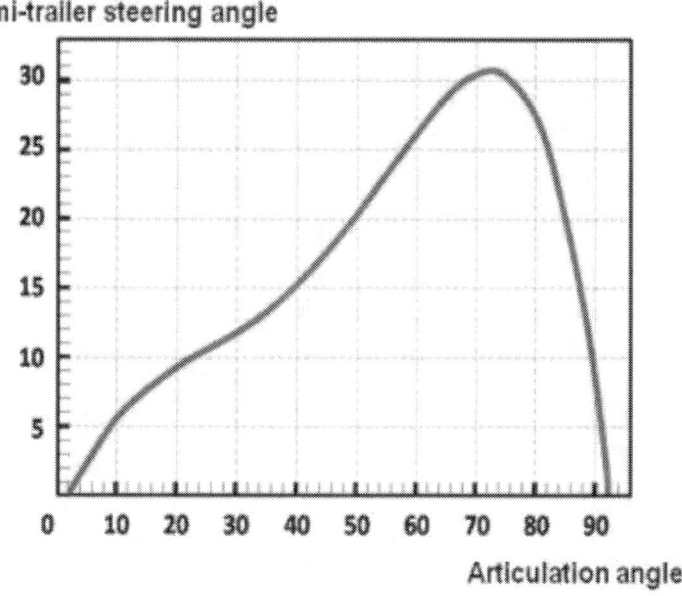

Figure 8. Articulation angle vs. steering angle on semi-trailer axle to minimize off-tracking (values in degrees)

Figure 8 shows the final curve of response in the command steering axle of the semi-trailer in relation to the articulation angle. The resulting equation of control was defined as:

$$X=\exp(-1.370004G+0.099728H+0.833331J-1.102983) \qquad (1)$$

where

G = 5th wheel to rear tractor axle distance

H = 5th wheel pin to last trailer axle distance

J = spacing between trailer axles

Note: G, H and L are illustrated in figure 1.

$$Y=-3.36E-07LX^5+9.3698E-06LX^4+1.4672E-03LX^3-7.2697E-02LX^2+1.5551LX-1.9183 \qquad (2)$$

where

L = articulation angle

Factor X in equation 1 refers to the dimensional specification of the semi-trailer G, H, L and Y are all illustrated in Figure 1. In equation 2, L is the articulation on the fifth wheel measured in degrees and Y corresponds to the angle required by the steering axle of the semi-trailer to minimize off-tracking.

Fuzzy-logic control

As well as prior prototypes developed in Australia and England, endow of steering control in the semi-trailer improves maneuverability, but also increases the risk of vehicular instability.

During the testing, there was some experimentation with certain parameters that could minimize potential stability loss, for example, consider vehicle's speed. These defined behaviors would apply during maneuvers at lower vehicular speeds and, as speed is increased, the system's angle of the steering axle of the semi-trailer should decrease proportionally, to the point were the system will be set to null response at speeds higher than 60 Km/h (Figure 9).

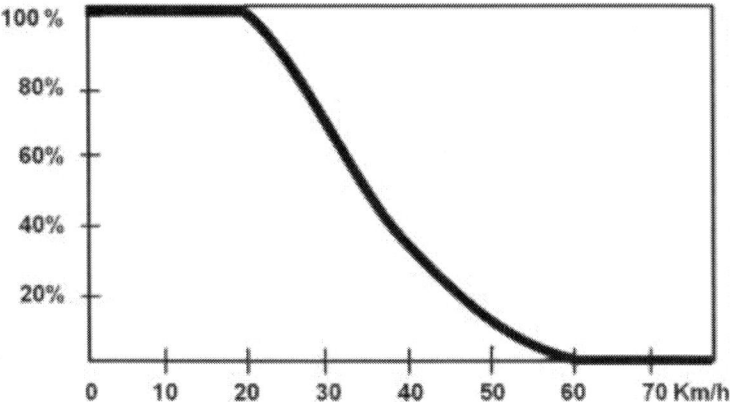

Figure 9. Response in the system based in vehicular velocity

Also the vehicle's speed has an impact on the response speed of the system which means that at very low vehicular speeds, the system will promptly execute angular adjustments, but as vehicle speed increases, these adjustments to the steering wheels will be performed more slowly.

Another parameter to be considered is the marginal angular changes. This consists of a time delay in the response of the system to changes in the articulation angle in order to prevent undesired oscillations that could occur during fast angle variations for example when the vehicle encounters a road or speed bump. If the angle variation doesn't remain for at least two seconds, the system should ignore it.

Finally, all this considerations will be affected by the weight carried in the semitrailer, due this and the multi-variable and non-linear nature of the system, an intelligent control technique was required, taking into consideration fuzzy logic which is part of artificial intelligence techniques and is an attractive and well-established approach to solving control problems (Lee, 1990).

Fuzzy logic provides a methodology to represent, manipulate and implement expert heuristic knowledge for controlling a system. Also, with a fuzzy controller, robustness and low cost are inherent to each development (Passino, 1998).

The set of fuzzy rules to develop the control system for each variable was obtained and fine tuned during the test rides of the experimental semi-trailer.

Figure 10 shows the example of two membership groups: a) articulation angle, with six linguistic variables to represent all the angle values: Zero, 4 to 20, 20 to 50, 50 to 74, 74 to 82 and 82 to 90; as an example 4 to 20 means that the value is between 4 and 20 degrees, considering zero degrees when the tractor and the semitrailer are aligned or in angles between 0 and 4°, and b) vehicular speed, with four linguistic variables: low, middle, middle-high, and high, both articulation angle and vehicular speed will be used to feed the fuzzy angle estimator in order to calculate the angle of the steering axles.

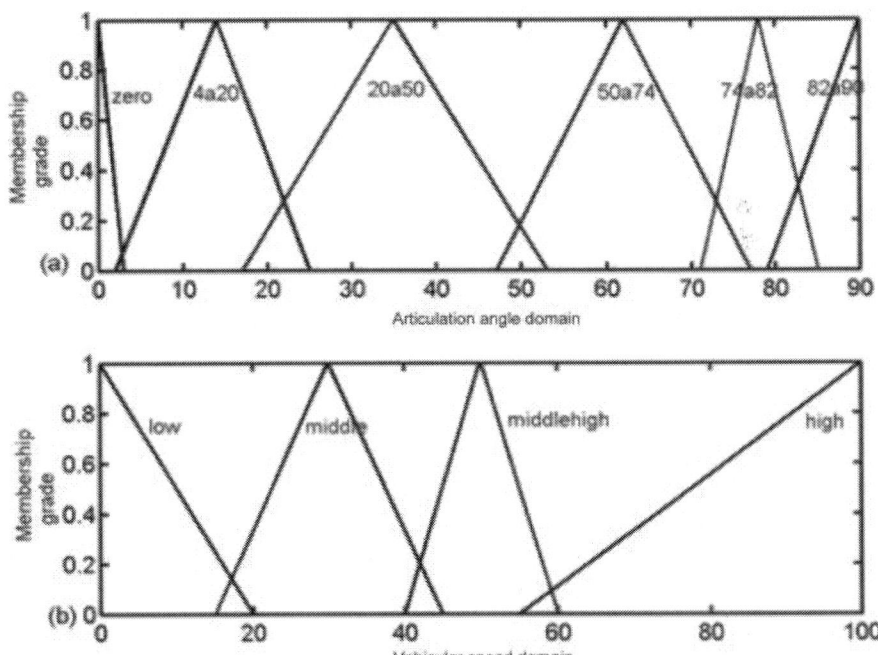

Figure 10. Membership functions: a) articulation angle and b) vehicle speed

This calculated angle is then compared with the current steering wheels angle to determine the corrections that should be applied to the steering mechanism (block diagram illustrated in Figure 11).

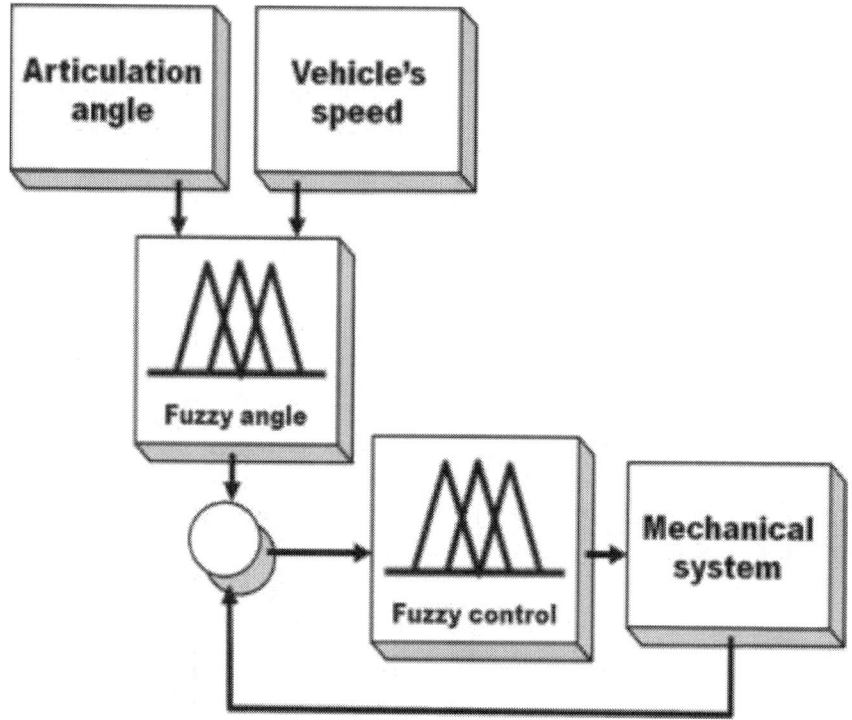

Figure 11. Fuzzy-controller block diagram

The corrections made by the fuzzy mechanical control are then compared cyclically retro-feeding the system to avoid over-steer. Figure 12 shows the membership functions group belonging to the error variable of the fuzzy control module, with "low", "ok" and "high" as linguistic variables, to deal with the vehicular specific characteristics, such as: deformations caused by the lateral forces, deformations caused by the longitudinal forces, deformations caused by the auto-alignment forces, semi-trailer length, rigidity of the steering mechanism and drift rigidity of the tires.

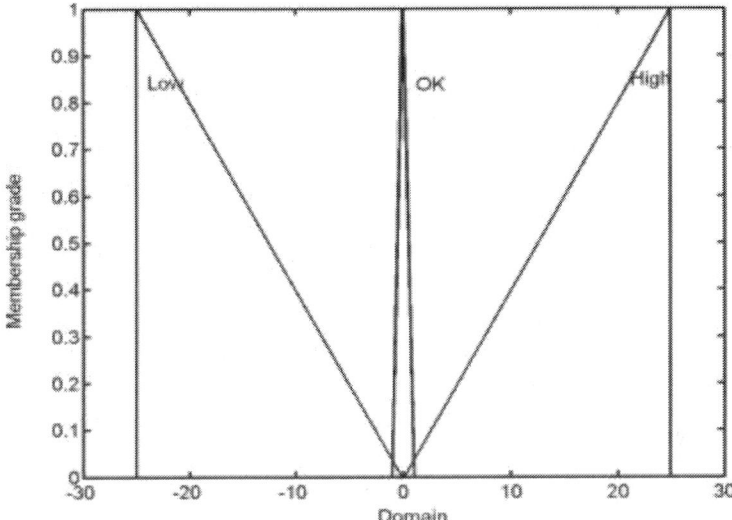

Figure 12. Error membership function

CONCLUSION

Heavy articulated vehicles are used world-wide, as one of the most feasible solutions for freight. Due to the length dimensions and weight they encounter poor maneuverability.

The handling of this kind of such vehicles can be greatly improved by using steering axles in the semitrailer to reduce the required road surface necessary to perform a turn maneuver placing less demand on lateral tire efforts and consequently decreasing the risk of loosing vehicular directional control, not to mention the reduction of tire wear and scrub. Likewise, by reducing lateral tire friction, the damage to road infrastructures will be minimized.

Fuzzy control represents a useful tool in dealing with non-linear systems. The controller approach presented in this paper is an alternative to solve some of the problems related to command steering in semi-trailers without detriment to the improved maneuverability provided from this type of systems.

The growing demand in freight transportation systems will soon require more efficient vehicles that are more secure and intelligent, as well as environmentally friendly. IE: Will not cause damage to the road infrastructures. In the years to come these varied demands will lead to the implementation of new technologies and designs.

ACKNOWLEDGMENTS

This research was sponsored by Consejo Nacional de Ciencia, Tecnología, Universidad Autónoma de Querétaro and Instituto Mexicano del Transporte.

REFERENCES

1. Battelle Team Comprehensive Truck Size and Weight Study (1995) Federal Highway Administration, US Department of Transportation
2. J. Billing, J. Patten Performance of Infrastructure-Friendly VehiclesCenter for Surface Transportation Technology, National Research Council of Canada, Technical Report CSTT-HVC-TR-058, Canadá (2003)
3. P. Bolzern, A. Locatelli A Comparative Study of Different Solutions to the Path-tracking Problem of an Articulated Vehicle, Politecnico di Milano, IEEE International Conference on Control Applications (2002) UK
4. L. Bortoni, M. Martínez, G. Herrera, R. Castañeda On-Board Instrumentation to Assess Articulated Vehicle Maneuverability, included in Commercial Vehicle Advancements in BrakingSteering Systems and Vehicle Stability Control Affects, USA (2007)
    A. El-Gindy Comparison of Various Computer Simulation Models for Predicting the Directional Responses of Articulated VehiclesVehicle System Dynamics, Canada (1987), pp. 249–268
5. P. Hingwe, J. Wang, M. Tai, M. Tomizuka Lateral Control of Heavy Duty Vehicles for Automated Highway System: Experimental Study on a Tractor Semi-TrailerUniversity of California, Berkeley (2000)
6. B. Jujnovich, D. Cebon Comparative Performance of Semi-Trailer Steering Systems, on: 7th International Symposium on Heavy Vehicle

Weights and DimensionsUniversity of Cambridge, UK (2002) [on line], Available: www.cvdc.org

7. C.C. Lee Fuzzy Logic in Control Systems: Fuzzy Logic Controller, parts I and II

8. Transactions on Systems, Man, and Cybernetics, 20 (1990), pp. 404–432

9. S. Neads Path Following in Articulated VehicleAntony Best Dynamics, UK (2006)

10. M. Passino Fuzzy Control. Department of Electrical EngineeringAddison-Wesley, Longman Inc, California, USA (1998)

11. H. Prem, E. Ramsay Performance Evaluation of the Trackaxle Self-Steering SystemRTDynamics, Australia (2001)

12. M. Sampei, T. Tamura, T. Kobayashi, N. Shibui Arbitrary Path Tracking Control of Articulte Vehicules Using Nonlinear Control Theory IEEE Transactions on Control Systems Technology, 3 (1) (1995)

13. D. Sampson Active Roll Control of Articulated Heavy VehiclesChurchill College, University of Cambridge, UK (2000)

14. S. Sankar, S. Rakheja, A. Piche Directional Dynamics of a Tractor-Semitrailer with Self- and Forced-Steering AxlesConcave Research Centre, Concordia University, Canada (1991)

15. J. Tsao, Y. Dessouky, K. Rangavajhula, J. Zeta, L. Zhou Automatic Steering For Conventional Truck Trailers (2006) San Jose State University, California PATH Research Report, USA

CITATION

L. Bortoni-Anzures, D. Gómez-Meléndez, G. Herrera-Ruíz, M. Martínez-Madrid, Fuzzy Controller for Automatic Steering in Heavy Vehicle Semi-Trailers, Ingeniería, Investigación y Tecnología, Volume 14, Issue 1, January–March 2013, Pages 1-9, ISSN 1405-7743, http://dx.doi.org/10.1016/S1405-7743(13)72220-0.

CHAPTER 8

A Sensor System for Online Oil Condition Monitoring of Operating Components

Manfred R. Mauntz[1], Jürgen Gegner[2], Ulrich Kuipers[3] and Stefan Klingau[1]

[1]cmc Instruments GmbH, Eschborn, Germany

[2]University of Siegen, Siegen, Germany

[3]South Westphalia University of Applied Sciences, Hagen, Germany

INTRODUCTION

A web-based oil diagnosis system for continuous online lubricant condition monitoring is presented. The new approach utilizes sensor detection of chemical aging of engineering oils and their additives or first traces of wear debris by precision measurement of the electrical properties. The basic concept and physical background are introduced in detail.

The application potential of the sensor system is discussed on the example of the early identification of critical operating conditions for premature white etching cracks failures of rolling bearings in industrial gearboxes. Causative vibration loading is revealed prior to any component damage. Large roller bearings in wind energy gearboxes unusually often fail prematurely, i.e. clearly before the nominal L_{10}life. The failure is characterized by axial raceway cracks, from which branching and spreading crack systems, partly decorated by white etching microstructure, develop into the depth by corrosion fatigue. High localized friction coefficients, resulting from the specific vibration caused mixed friction operating conditions, initiate tensile stress induced cleavage-like brittle spontaneous surface cracking. The basic idea of the novel failure detection condition monitoring system is the early identification of chemical aging of the lubricant and its additives under the influence of vibration loading.

The sensor effectively controls the proper operation conditions of, e.g., bearings and gears in gearboxes. The online diagnostics system measures components of the specific complex impedance of oils. For instance, metal abrasion due to wear debris, broken oil molecules, forming acids or oil soaps, result in an increase of the electrical conductivity, which directly correlates with the degree of contamination of the oil. For additivated lubricants, the stage of degradation of the additives can also be derived from changes in the dielectric constant. The determination of the reduction in the oil quality by contaminations and the quasi continuous evaluation of wear and chemical aging follow the holistic approach of a real-time monitoring of an alteration in the condition of the oil-machine system. The measuring signals can be transmitted online to a web-based monitoring system via LAN, WLAN or serial interfaces of the sensor. Control of the relevant damage mechanisms, e.g. tribiological wear or oil aging, during proper operation below certain tolerance limits then allows preventive, condition-oriented maintenance to be carried out, if necessary, long before regular overhauling. Outage durations are reduced and the life of components and machines is increased.

BASIC SENSOR CONCEPT AND PHYSICAL PRINCIPLES

Basic Sensor Concept

The basic sensor concept of the novel engineering oil monitoring system is based on the measurement of complex oil impedance components X, particularly the specific electrical conductivity κ and the relative permittivity ε_r. Due to their temperature dependence, the oil temperature T is also recorded [1-3]. Two or more electrodes, between which the oil flows, serve as a basic sensor. Resistance and capacity are measured independently of each other. Zero-mean periodic quantities are used to prevent polarization effects. Figure 1 shows the sensor in its triple plate design.

Figure 1. Sensor in triple plate design.

Oils are electrical non-conductors. The electrical residual conductivity of pure oils lies in the range below 1 pS/m. For comparison, the electrical conductivity of the electrical non-conductor distilled water is larger by six orders of magnitude.

Broken oil molecules, acids, abrasive (metallic) wear, ions, oil soaps, etc., cause an increase of the oil conductivity κ. It rises with increasing ion concentration and mobility. The electrical conductivity of almost all impurities is high compared to the extremely low corresponding property of original pure oils.

The basic sensor represents an electrode arrangement, in which the measured oil is used as electrical conductor and as dielectric material for conductivity and relative permittivity measurement, respectively. Oil is an electrical non-conductor. High resistance of the basic sensor and resulting low measurement currents provide best interference sensitivity to interspersed electromagnetic fields. Due to the very small currents, moreover, sufficient interference suppression is achieved. To prevent polarization effects, zero-mean alternating current voltages are measured as test signals. However, no capacitive current components may be measured simultaneously during a conductivity measurement because the capacitive current is much higher than its ohmic components. Thus, rather high requirements are set on analog sensor electronic systems, which are met with the reported measurement procedure.

The conductivities of the insulating construction elements and insulation of electrical feedthrough are about the same size as for the pure oils to be analyzed. The developed basic sensors and precise sensor electronic system ensure that the conductivity of feedthroughs and substrates may not be included into the test results. The active basic sensor unit consists of two or several basic sensor plates which are fixed to metal pins of a glass/metal feedthrough in a constant distant from each other. The plates of the basic sensor are arranged in the middle of the measuring chamber, allowing for an adequate incident flow of the flowing medium. A special alignment of the sensor housing parts is thus not necessary in this design. The extension characteristics of the sensor housing materials and the glass/metal feedthrough pins are exactly adjusted to the material characteristics of the used feedthrough glass. The compression strength is above 10 MPa

Temperature Compensation
The ion mobility and thus the electrical conductivity κ depend upon the internal friction of the oil and therefore also on its temperature. The oil conductivity increases with temperature. Figure 2 shows the dependence of the conductivity κ on the temperature change ΔT.

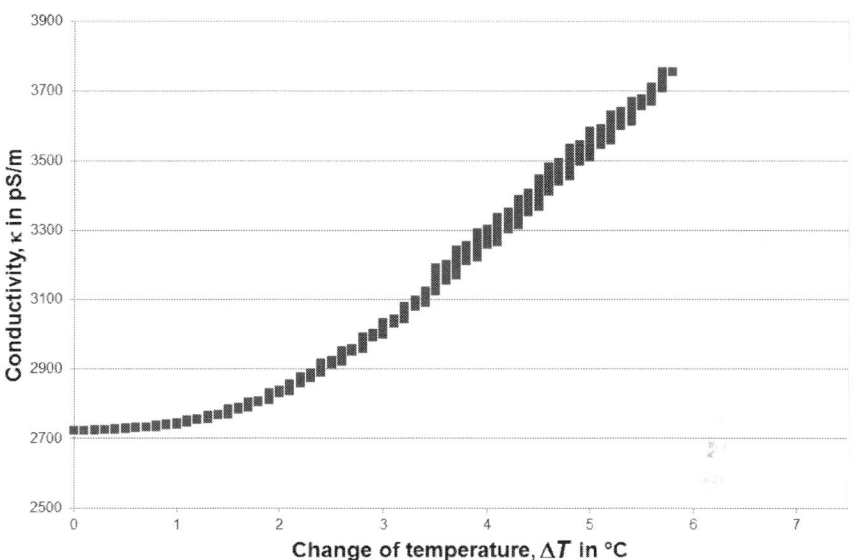

Figure 2. Temperature dependence of the electrical conductivity of sample oil.

Already for about 3 °C alteration in temperature, the conductivity changes by about 25%. The electrical conductivity κ is a temperature function that depends on oil impurities rather than on the oil itself. The type of pollution and its temperature dependence cannot be assumed to be known. To improve the comparability of measurements, a self-learning adaptive temperature compensation algorithm is implemented. An integral alteration of the oil quality can then be assessed by the temperature compensated conductivity value, whereas the type of contamination is not determinable. The relative permittivity is measured with the same basic sensor arrangement as used for the electrical conductivity.

The electrical conductivity and relative permittivity are to be measured with respect to a reference temperature T_R as close as possible to the operating temperature of the oil. These parameters can be evaluated by means of temperature-dependent approximating polynomials, as demonstrated below exemplarily for the electrical conductivity:

$$\kappa_R = \kappa_{R,0} + \left(a\Delta T_C + b\Delta T_C^2 + c\Delta T_C^3\right) \times \kappa_M$$

(1)

Here, κ_R and $\kappa_{R,0}$ denote the approximate and previously calculated (old) electrical conductivity of the oil at the reference temperature T_R, respectively. T_C stands for the current temperature of the oil and κ_M is the electrical conductivity measured without temperature compensation. Moreover, a, b, und c are the coefficients of the approximating polynomial to be adaptively determined. The temperature difference is defined as follows:

$$\Delta T = T_R - T_C$$

(2)

The oil temperature T_C is measured for this temperature compensation. The use of a polynomial of the third order in Eq. (1) ensures good approximation while keeping the computational effort for the applied microcomputer reasonably low. Figure 3 shows the measured values of the electrical conductivity κ after temperature compensation.

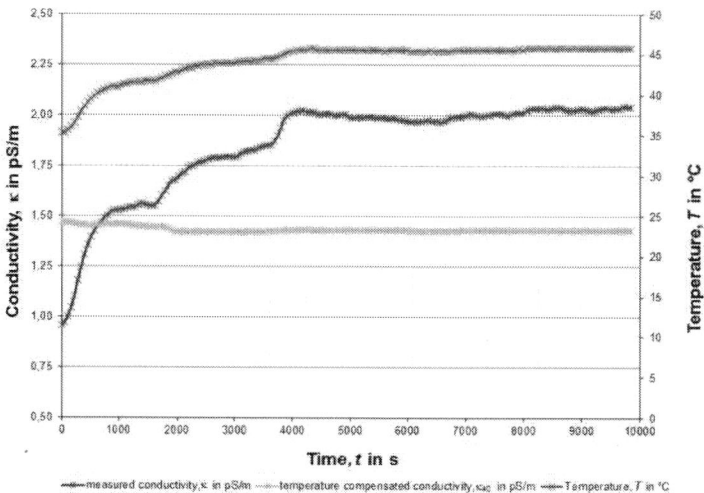

Figure 3. Measured conductivity values after temperature compensation.

Calculation and Linear Approximation of Relative Permittivity and Conductivity

In a series of experiments on the non-additivated lubricating oil FVA03, fresh demineralized water was added to a volume of 3.01%. The oil conductivity data measured as a function of the water content are found to follow a linear relationship in good approximation. The theoretical course of the relative permittivity is calculated for dilute solutions according to different mixing rules by truncating a Taylor series expansion of the model equations after the linear term. The model of Lichtenecker is evaluated inFigure 4. Lichtenecker developed the formula of Eq. (3) for calculating the dielectric constant of a homogeneous mixture ε_r [4]:

$$\varepsilon_r = \varepsilon_{r,add}^f \times \varepsilon_{r,oil}^{1-f}$$

(3)

The permittivity of the addition and the oil, respectively, is denoted $\varepsilon_{r,add}$ and $\varepsilon_{r,oil}$. With the volume fraction f of the addition, $1-f$ becomes the volume fraction of the oil.

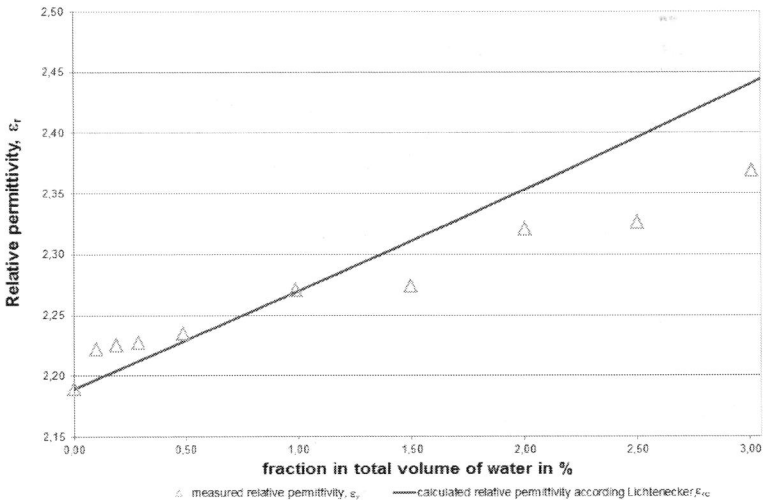

Figure 4. Electrical permittivity ε_r measured as a function of the water content and model fit according to Eq. (3).

PREMATURE FAILURES OF ROLLING BEARINGS AND CORRELATION WITH OIL AGING

Bearings in industrial, e.g. wind turbine, gearboxes unusually often suffer from a significantly shorter life than calculated by white etching cracks [5, 6]. Figure 5 shows the light-optical micrograph of a typical metallographic microsection [5]. The overrolling direction from left to right indicates surface initiation and top-down propagation of the extended crack system.

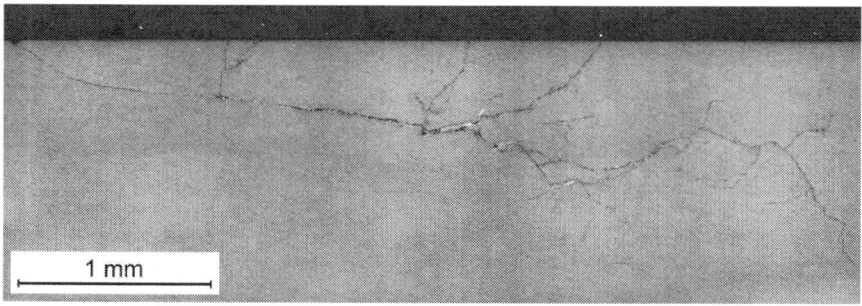

FIGURE 5. Radial microsection of a branching and spreading white etching crack system.

These early failures are characterized by mostly axial raceway cracks, revealing vertical semi to fully circular cleavage-like lenticular brittle spontaneous incipient cracks in preparatively opened original fracture faces [5, 6]. Occasionally, pock-like spallings are associated with the surface cracks, as shown exemplarily in Figure 6 [6]. The developing deep crack systems are open to the raceway, from which oil penetrates and promotes further corrosion fatigue crack growth as well as local secondary microstructural changes in the form of crack path decorating white etching constituents. It is evident from fractography and X-ray diffraction (XRD) residual stress analyses that the cleavage-like incipient cracks are caused by frictional tangential tensile stresses

[5, 6], which occur in subregions of the contact area in specific, vibrationally induced mixed friction operating conditions [5–8].

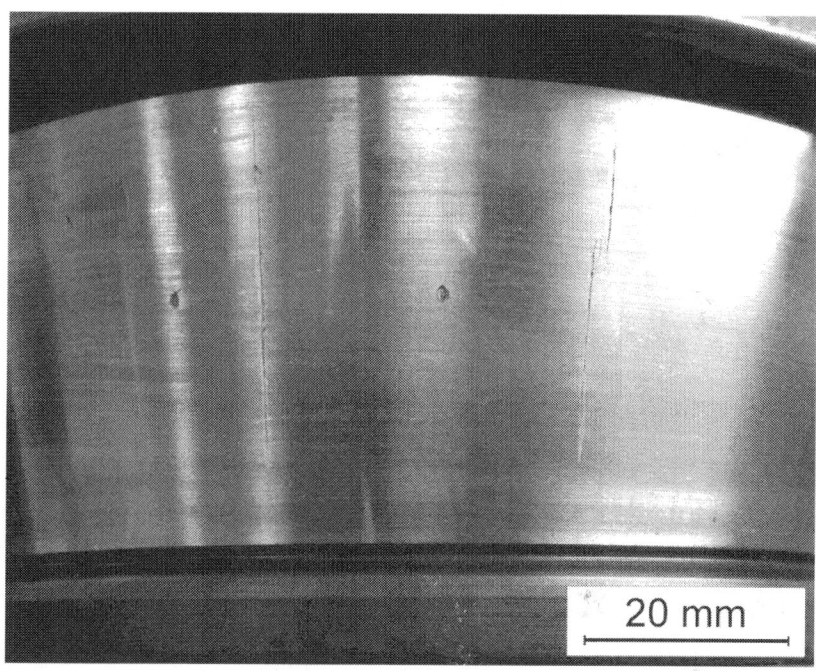

Figure 6. Inner ring raceway with typical axial cracks and few associated pock-like spallings.

XRD based material response analyses of run rolling bearings, suffering from white etching cracks on still largely undamaged raceways, reveal the causative vibration loading [5]. It is further reported that lubricant aging occurs under the influence of vibrations [7, 8]. An example of an infrared spectrum of used oil from rig test run of a roller bearing is provided in Figure 7 [7]. The verified O–H and C=O oxidation bands indicate operational acidification of the oil, also reflected in the dissolution of MnS inclusion lines on the raceway (cf. Figure 8), as a result of polycondensation reactions towards resinification and beginning lacquer formation. It is this aging of the lubricating oil and its additives, which can be detected at an early stage by the new sensor so that a gearbox operating at critical conditions is identified.

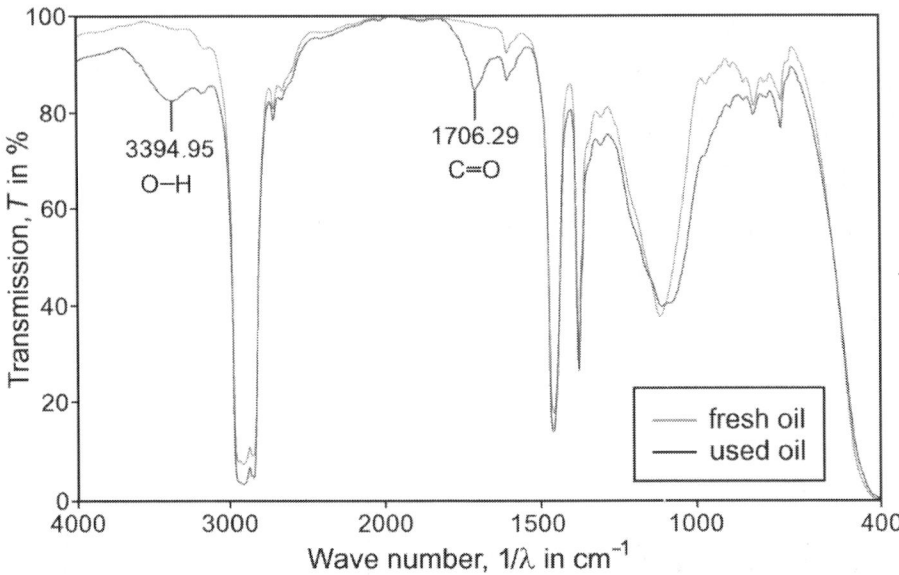

Figure 7. Oxidation peaks in the infrared spectrum of a used non-additivated aliphatic lubricating oil run under vibration loading in a rolling bearing rig test (λ is the wavelength).

The mentioned crack initiation by tribochemical reactions is also found on lateral surfaces of rollers. InFigure 8 [5], a scanning electron microscope (SEM) image, taken in the secondary electron imaging mode, is shown on the left. Residues of manganese and sulphur, detected in the crack-like defect by energy dispersive X-ray spectroscopy (Figure 8, on the right), indicate the causative tribo chemical dissolution of nonmetallic MnS inclusions [5, 7, 8].

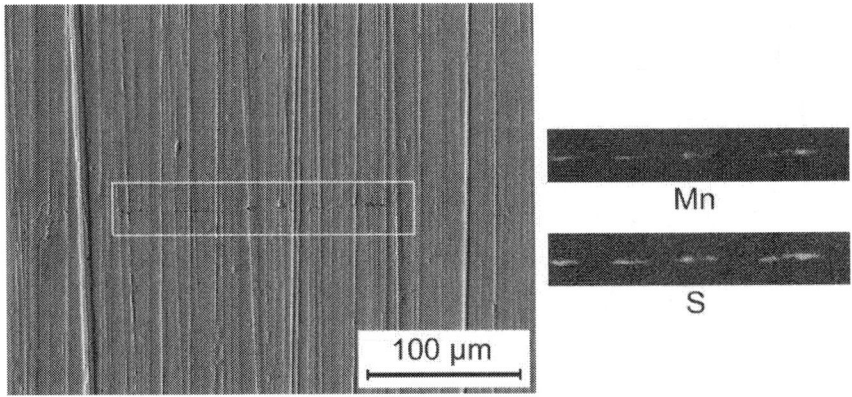

Figure 8. SEM image of a crack on a bearing roller with elemental mapping of Mn and S.

TRIAL OF THE OIL SENSOR SYSTEM ON A BEARING AND GEAR TEST RIG

On a bearing and gear test rig, the new sensor based oil quality monitoring system is applied. Various load cycles are run and speeds and torques are measured. The results of the trial are described, evaluated and discussed in the following sections.

Loss of Power and Trial Run Characteristics

The speed-related power $P(n)$ of the test rig is given as follows:

$$P = M \bullet \omega \; with \; \omega = 2\pi \bullet n$$

$$(4)$$

Here, M denotes the torque, ω and n respectively stand for the angular velocity and rotational speed. The implemented power loss ΔP is derived from the transmission ratio $N\ddot{U}$:

$$\Delta P = \omega 1 \cdot (M1 - N\ddot{U} \cdot M2)$$

$$(5)$$

M_1 and M_2 indicate the torque of the drive and the load, respectively. In the first trial on the test rig, the rolling bearing is intentionally damaged. The time-dependent power loss in the gear, as derived from the measuring signal characteristics, is represented in Figure 9.

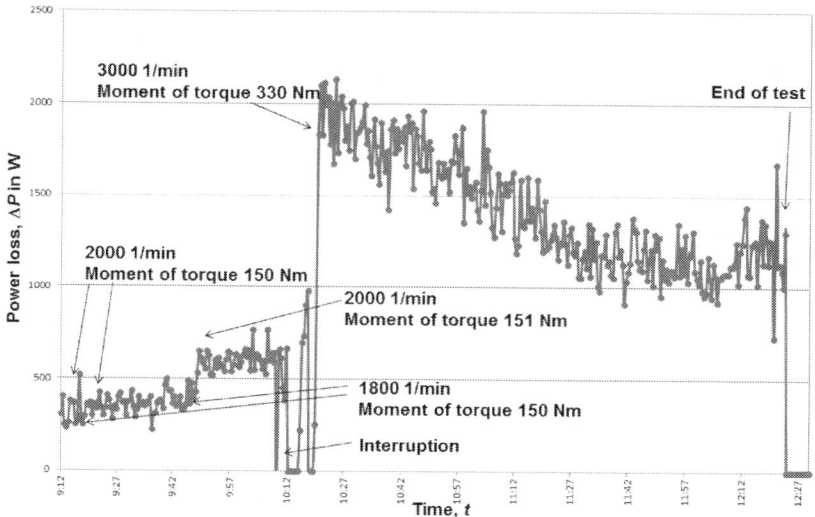

Figure 9. Calculated power loss ΔP.

After switching to a higher load, the power loss increases abruptly before the bearings run in. Towards the end of the trial, the bearings reveal indication of advanced deterioration. A wide lubrication gap and vibrations when overrolling spalling results in higher oscillation amplitudes, which leads to the automatic shutdown of the test rig eventually. The measuring results obtained with the oil sensor system are presented in the following.

Conductivity of The Lubricating Oil

Figure 10 shows the test readings of the conductivity measurement of the lubricating oil. The current bearing wear and the deteriorating oil condition in the conducted trial are reflected in the change of the electrical conductivity plotted vs. running time in the diagram.

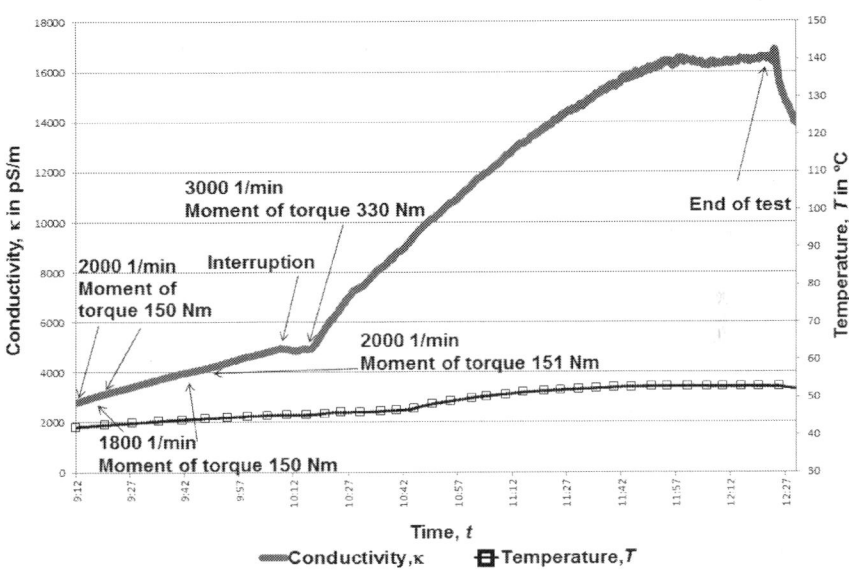

Figure 10. Measurement of the electrical conductivity κ vs. running time *t*.

New oil from the storage container exhibits a conductivity κ of 2312 pS/m. After filling into the trial gear and before the start-up of the test rig, a conductivity of 2791 pS/m is measured. This increase can be attributed to existing residual impurities in the gear. During the trial run, the conductivity κ of the gear oil increases to 16868 pS/m. Besides changes in temperature, conductivity increase is caused, e.g., by wear debris and removed material from spalling, impurities, broken oil molecules or forming oil soap. As described above, the temperature dependence of the electrical conductivity of the used gear oil is compensated and the oil conductivity measured in the gear trial is converted into the relevant conductivity value at 40 °C. Figure 11 shows the development of the temperature compensated oil conductivity with running time during the gear trial.

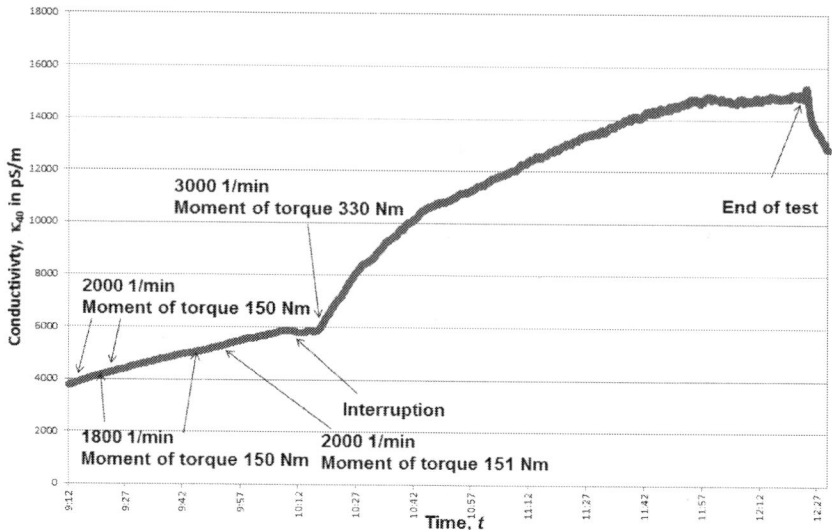

Figure 11. Time curve of the temperature compensated oil conductivity derived from Figure 10.

In the case of an initially low load, the electrical conductivity increases linearly with running time. It is to be assumed that the low bearing wear in this area also increases proportional to the time.

During the necessary intermediate shutdown (interruption) and run-up of the drive machine to 330 Nm, the conductivity is virtually constant. After switching over to the higher load, the oil conductivity increases strongly. Here, the bearing run-in (shakedown) is shown as reduction in the conductivity increase. More than about 30 minutes prior to the final forced shutdown of the trial run by an oscillation sensor, the conductivity remains almost constant followed by a temporary rise directly before disconnection. After switching off the test gear, the oil conductivity decreases strongly. This clearly emphasizes the influence of the additives. During the loading stages, more impurities per time unit are produced than bound to additives. After shutting down the test rig, no further oil contamination occurs while the effect of additives still continues.

The variations in electrical conductivity are depicted in Figure 12. In this diagram, the curve follows averages respectively calculated over 3 minutes.

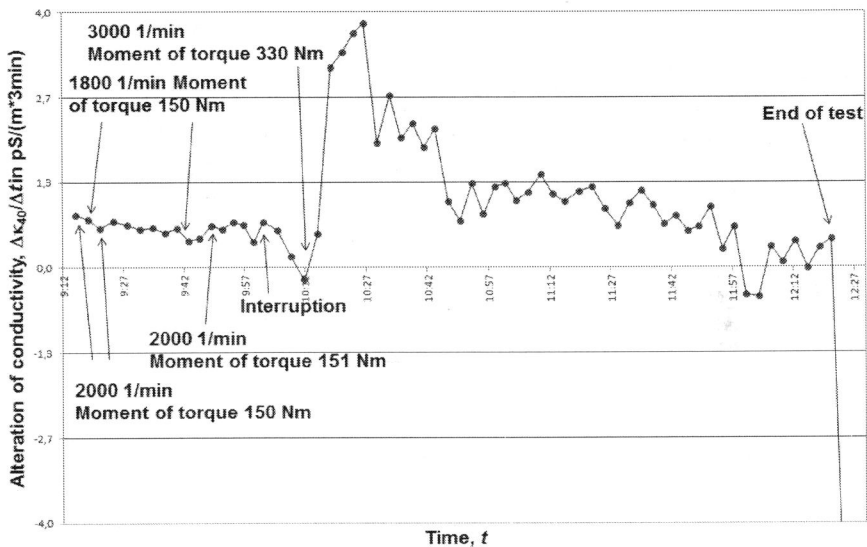

Figure 12. Alteration of the electrical conductivity, expressed as $\Delta\kappa_{40}/\Delta t$, vs. running time t.

When starting up at 2000 revolutions per minute and a torque of 150 Nm, a relatively constant change in conductivity from 0.6 to 0.8 pS/(m×3 min), equivalent to 3.3 to 4.4 fS/(m×s), occurs. In the case of higher load (330 Nm, 3000 min⁻¹), the change in conductivity rises up to 3.8 pS/(m×3 min), i.e. 21.1 fS/(m×s). After the intermediate load increase, the effect on the change of the oil conductivity appears stronger. This may be attributed to the time-dependent formation of impurities and changes in bearing stressing as can be expected during the development of spalling. Figure 13 shows the inner ring of the failed planet bearing with massive damage of the raceway at the end of the trial.

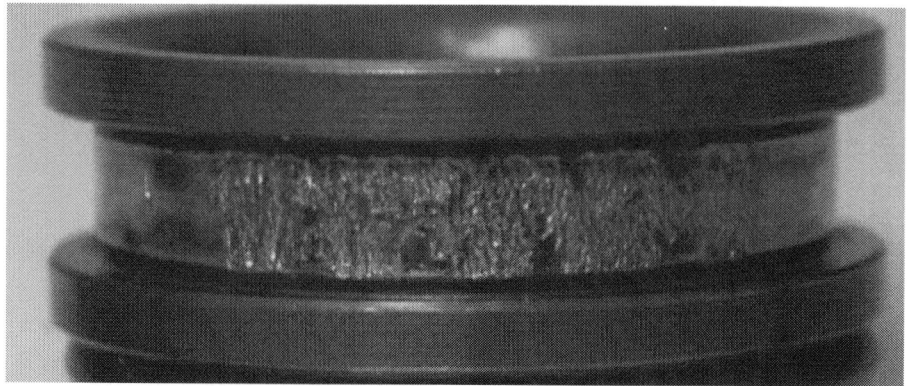

Figure 13. Heavily spalled inner ring raceway of the tested cylindrical roller bearing.

The connection between the change in conductivity κ and the loss of power ΔP in the gear is also evaluated. Figure 14 represents this progress graphically. Both the increasing change in oil conductivity and the gear power loss correlate with the bearing wear.

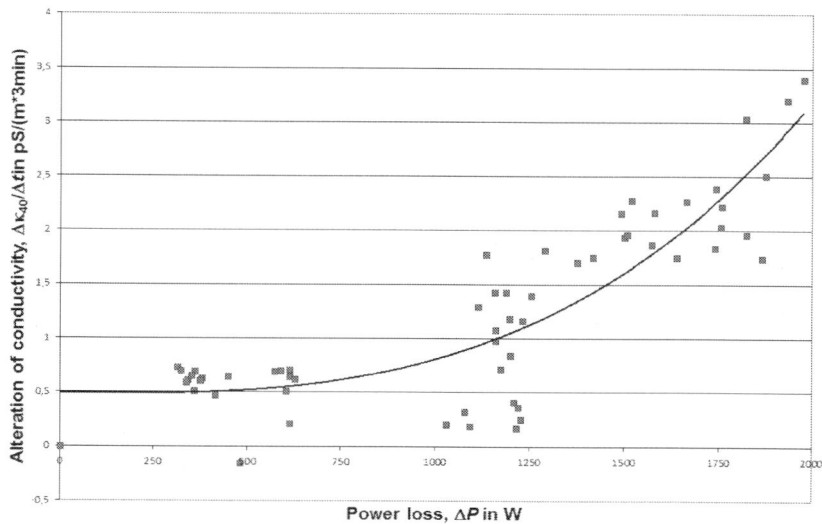

Figure 14. Alteration of the electrical conductivity as function of power loss of the gearbox.

In the diagram of Figure 14, a trend line is drawn as polynomial of the third order. The higher the increases in conductivity and loss of power in the gear, the stronger the bearing wear occurs. Exceptional changes to the system, e.g. switching of the load conditions, are not taken into account.

Relative Permittivity of the Lubricating Oil

In addition to the electrical conductivity, the relative permittivity ε_r of the oil is measured. In the case of oils not enriched with additives, the water content can be determined that way. There are good prospects that the dwindling efficacy of the additives can be detected by means of the dielectric constant measurement. Figure 15 shows the time development of the relative permittivity during the trial run. Due to the dependence on temperature, this development is also depicted in the diagram.

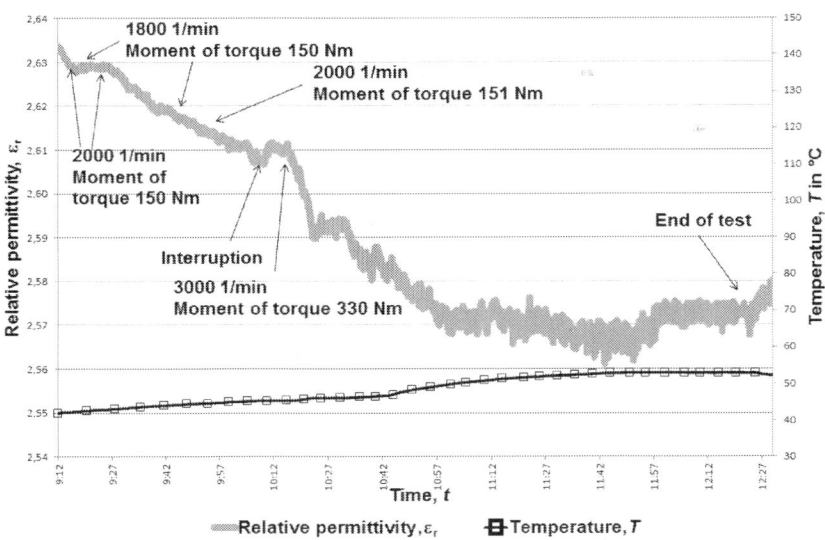

Figure 15. Time curves of the relative permittivity ε_r and the temperature T as a function of the running time t.

The change of the relative permittivity could be caused by a combination of the effects of a chemical reaction of additives, water evaporation from the oil and the temperature dependence of the cell constants as well as the relative permittivity itself. During the trial run, the temperature increases from 42 to 52.9 °C. The temperature dependence justifies the developed adaptive, self-learning temperature compensation technique. Figure 16 shows the temperature compensated time development of the relative permittivity during the gear trial.

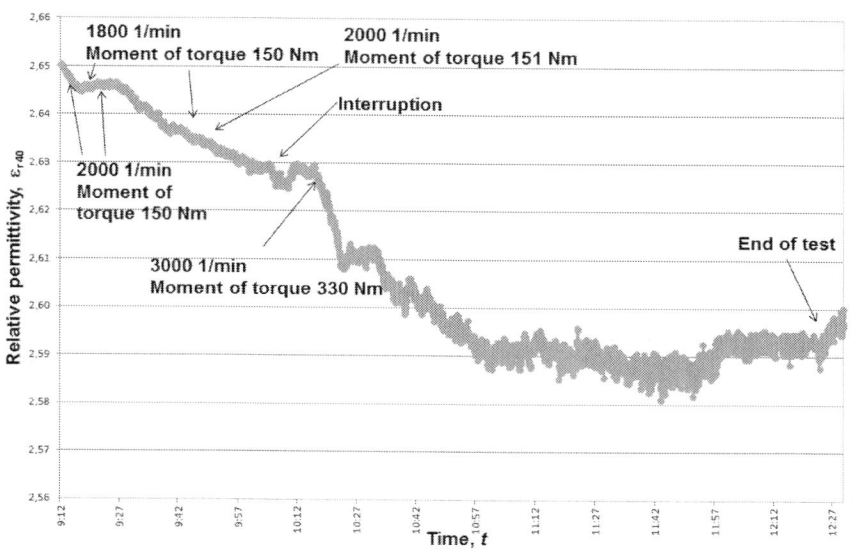

Figure 16. Time curve of the temperature compensated relative permittivity ε_r.

APPROACH FOR CONDITION MONITORING OF ADDITIVATED LUBRICATING OILS

A direct connection between the electrical conductivity and the degree of contamination of oils is found. An increase of the electrical conductivity of the oil in operation can thus be interpreted as increasing wear or contamination of the lubricant. The aging of the oil is also evident in the degradation of additives. The used additives reveal high conductivity compared with the oil.

The consumption of the additives is reflected in a reduction of the electrical conductivity and permittivity of the oil. The gradient, i.e. the time derivative, of the conductivity or the dielectric constant progression respectively represents a measure of the additive degradation and consumption. The full additive degradation is indicated by the slope of zero (bathtub curve). Then the measurement signal increases further with increasing pollution, water entry, etc. Figure 17 schematically shows the temperature compensated time curve of the permittivity of additivated oil continuously contaminated by the addition of wear debris, water or oil acids from chemical aging. Once the additives are consumed, the vanishing shielding effect results in a characteristic re-increase.

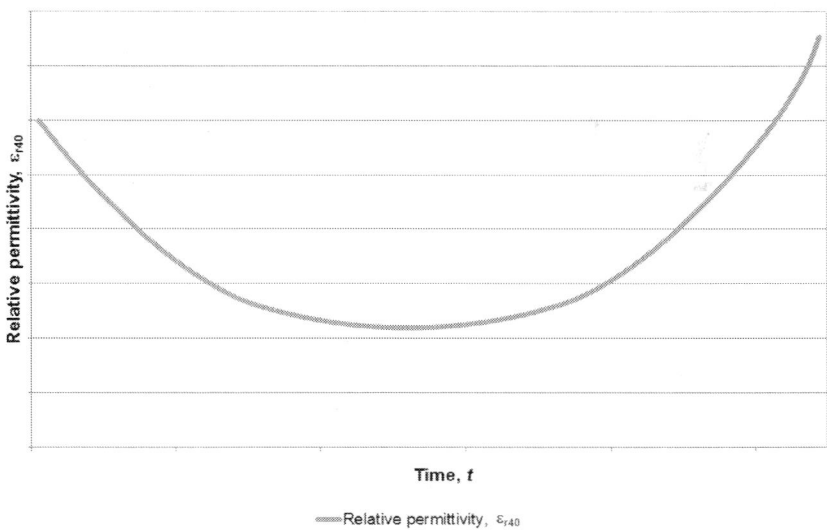

Figure 17. Temperature compensated permittivity.

The most commercially available particle counters only detect particles as small as 4 µm. In a very early stage of wear of bearings, gears, hydraulic cylinders, etc., however, particularly smaller particles are produced. A preventive maintenance lowing, rather than rigid inspection intervals, therefore requires recognition of even the smallest particles. These particles are far more common in the oils of functioning machines than larger ones. Oil aging can be involved in the failure, for instance, of rolling bearings [7].

WEB-BASED DECENTRALIZED LUBRICANT QUALITY MONITORING SYSTEM

The integration into a suitable communication structure and the realization of an online monitoring system offers an interesting practice-oriented utilization of the oil sensor system. This is briefly discussed below.

Preferred areas of application of the sensor system are energy production and automated technical plants that are operated locally, like e.g. wind turbines, generators, hydraulic systems or gearboxes. Plant employers are interested in continuous automated in vivo examination of the oil quality rather than interrupting the operation for regular sampling. Online oil status monitoring significantly improves the economic and ecological efficiency by increasing operating safety, reducing down times or adjusting oil change intervals to actual requirements. Once the oil condition monitoring sensors are installed on the plants, the measuring data can be displayed and evaluated elsewhere. A flexible decentralized monitoring system also enables the analysis of measuring signals and monitoring of the plants by external providers. A user-orientated service ensuring the quantitative evaluation of changes in the oil-machine system, including the recommendation of resulting preventive maintenance measures, relieves plant operators, increases reliability and saves costs.

In a web-based decentralized online oil condition monitoring system, the sensor signals are preferably transferred through the Internet to a database server and recorded on an HTML page as user interface [8]. Figure 18 shows the displayed measured data.

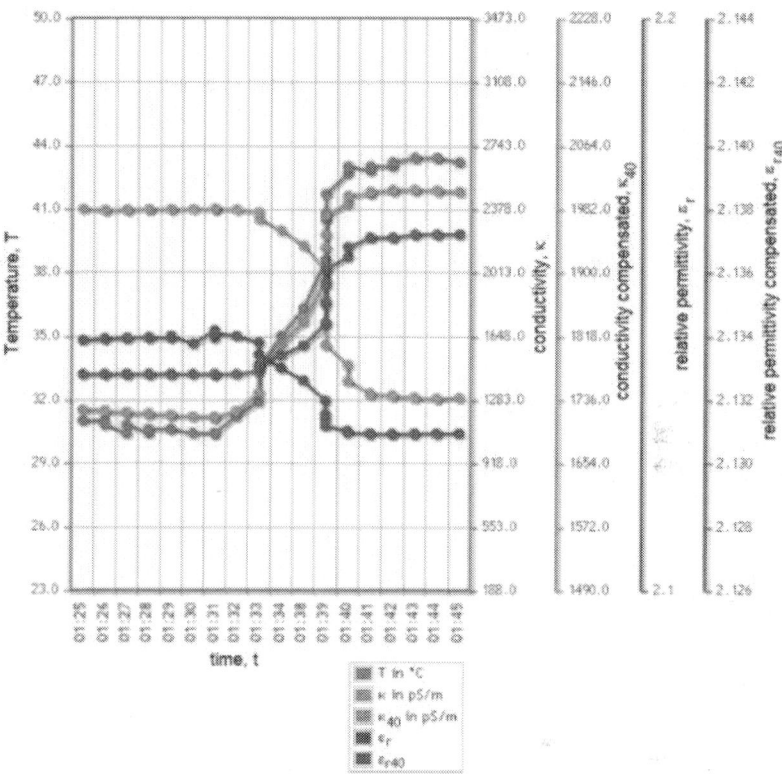

Figure 18. Displayed measured data.

Following authentication, a simple web browser permits access via the wired or wireless LAN. In case of alarm signals, an immediate automated generation of warning messages, for instance by e-mail or SMS, is possible from any computer with Internet connection. Figure 19 shows the new sensor system with communication unit [9].

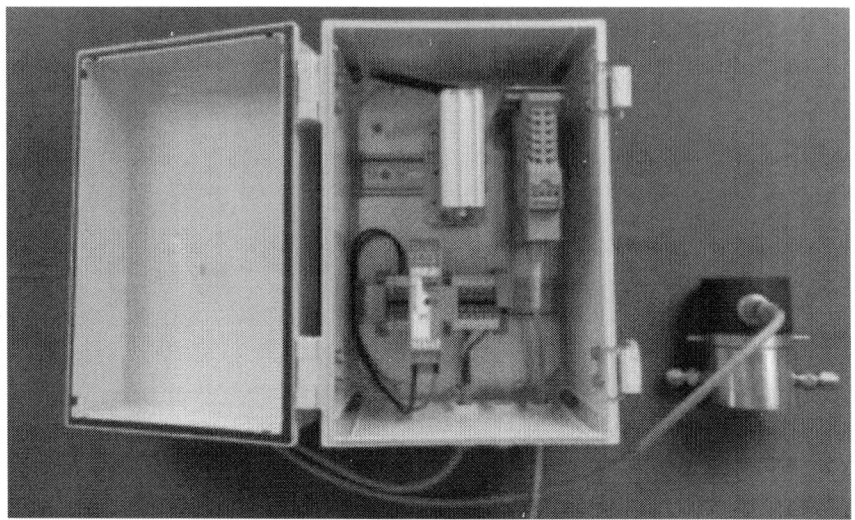

Figure 19. WearSens® sensor system with communication unit.

CONCLUSIONS

The online diagnostics system measures components of the specific complex impedance of oils. For instance, metal abrasion due to bearing wear at the tribological contact, broken oil molecules, acids or oil soap cause an increase in electrical conductivity that directly correlates with the degree of pollution of the oil. The dielectrical properties of the oils are especially determined by the water content, which, in the case of products that are not enriched with additives, becomes accessible by an additional accurate measurement of the relative permittivity. In the case of oils enriched with additives, statements on the degradation of additives can also be deduced from recorded changes in the relative permittivity.

Indication of damage and wear is measured as an integral factor of, e.g., the degree of pollution, oil aging and acidification, water content and the decomposition state of additives or abrasion of the bearings. It provides informative data on lubricant aging and material loading as well as the wear of the bearings and gears for the online operative monitoring of components of machines.

Additional loading, for instance, by vibration induced mixed friction in rolling-sliding contact (rolling bearings, gears, cams, etc.) causes specific faster oil aging, e.g., in the course of premature component failures. Verified in roller bearing vibration rig tests, the oil suffers from significant acidification by polycondensation reactions and incipient resinification, as proven by infrared spectroscopy of used lubricant. The application potential of the sensor is discussed on the example of the prevention of early rolling bearing failures in industrial gearboxes, of which vibrational contact loading is the root cause.

For an efficient machine utilization and targeted damage prevention, the new electrical online oil condition monitoring system offers the prospect to carry out timely preventative maintenance on demand rather than in rigid inspection intervals. The determination of impurities or reduction in the quality of the lubricants and the quasi continuous evaluation of the bearing and gear wear and oil aging meet the holistic approach of a real-time monitoring of a change in the condition of the oil-machine system.

The measuring signals can be transmitted to a web-based condition monitoring system via LAN, WLAN or serial interfaces of the sensor system. The monitoring of the tribological wear mechanisms during proper operation below the tolerance limits of the components then allows preventive, condition-oriented maintenance to be carried out, if necessary, long before regular overhauling, thus reducing outages caused by wear while simultaneously increasing the overall lifetime of the oil-machine system.

On a bearing and gear rig test, various load cycles are run and the functionality of the introduced electric online condition monitoring sensor system is tested successfully. The evaluation of the experiment is presented.

REFERENCES

1. Gegner, J., Kuipers, U., Mauntz, M. Ölsensorsystem zur Echtzeit-Zustandsüberwachung von technischen Anlagen und Maschinen, Technisches Messen 77; 2010. pp. 283-292.

2. Kuipers, U., Mauntz, M. Ölsensorsystem – Sensorsystem zur Messung von Komponenten der komplexen Impedanz elektrisch gering leitender und nichtleitender Fluide, dessen Realisierung und Anwendung, German Patent Application N° 10 2008 047 366.9, Applicant: cmc Instruments GmbH, German Patent Office, Munich, Filing date: 15.09.2008, in German.

3. Kuipers, U., Mauntz, M. Verfahren, Schaltungsanordnung, Sensor zur Messung physikalischer Größen in Fluiden sowie deren Verwendung, European Patent Application N° EP 09000244, European Patent Office, Munich; 09.01.2009, in German.

4. Lichtenecker, K., Rother, K. Die Herleitung der logarithmischen Mischungsgesetzes aus allgemeinen Prinzipien der stationären Strömung, Physikalische Zeitschrift, 1931, 32, pp. 255-260.

5. Gegner, J. Tribological Aspects of Rolling Bearing Failures, In: C.-H. Kuo (ed.), Tribology – Lubricants and Lubrication, Rijeka: InTech; 2011. Chap. 2, pp. 33-94.

6. Nierlich, W., Gegner, J. Einführung der Normalspannungshypothese für Mischreibung im Wälz-Gleitkontakt, Düsseldorf: VDI Reports 2147, VDI Wissensforum; 2011. pp. 277-290, in German.

7. Gegner, J., Nierlich, W. Operational Residual Stress Formation in Vibration-Loaded Rolling Contact, Advances in X-ray Analysis, Vol. 52; 2008, pp. 722-731.

8. Nierlich, W., Gegner, J. Material Response Bearing Testing under Vibration Loading, In: J. M. Beswick (ed.), Advances in Rolling Contact Fatigue Strength Testing and Related Substitute Technologies, STP 1548, ASTM International, West Conshohocken, Pennsylvania, USA, 2012.

9. Gegner, J., Kuipers U., Mauntz, M. New Electric Online Oil Condition Monitoring Sensor – an Innovation in Early Failure Detection of Industrial Gears, 4th International Multi-Conference on Engineering and Technological Innovation, 19.-22-07.2011, Orlando, Florida, USA 2011, Proceedings Volume I, International Institute of Informatics and Systemics, Winter Garten, Florida, USA, 2011, pp. 238-242.

10. Gegner, J., Kuipers U., Mauntz, M. High-precision online sensor condition monitoring of industrial oils in service for the early detection of contamination and chemical aging, Sensor+Test Conferences 07.-09.06.2011, Nürnberg, AMA Service GmbH, Wunstorf, 2011, pp. 702-709.

CITATION

Manfred R. Mauntz, Jürgen Gegner, Ulrich Kuipers and Stefan, Klingau (2013). A Sensor System for Online Oil Condition Monitoring of Operating Components, Tribology - Fundamentals and Advancements, Dr. Jürgen Gegner (Ed.), ISBN: 978-953-51-1135-1, InTech, DOI: 10.5772/55737.

CHAPTER 9

System Safety of Gas Turbines: Hierarchical Fuzzy Markov Modelling

G. G. Kulikov[1], V. Yu. Arkov[1] and A.I. Abdulnagimov[1]

[1]Automated Control and Management Systems Department, Ufa State Aviation Technical University, Ufa, Russia

INTRODUCTION

Reliability, safety and durability represent important properties of modern aircraft, which is necessary for its effective in-service use.

The reason of the main hazard for aircraft are both random and determined negative influences rendering the controlled object during its use. Faults, failures, disturbances, noises, influences of environment and control errors represent the objectively existing stream of random negative influences on the object.

Statistically, in the recent years the majority of aircraft incidents are connected with the human factor and late fault detection in plane systems. In this regard, requirements to flight safety which demand development of new methods and algorithms of control-and-condition monitoring/ diagnostic for complex objects raise every year. The analysis of modern gas turbine engines has shown that most faults appears in the engine itself and its FADEC (40-75% for FADEC, Figure 1).

The percentage of faults for FADEC depends on the achieved values for no-failure operation indicators of the engine and FADEC.

During the development of FADEC, it is necessary to adhere to the principles and methods guaranteeing safety and reliability of aircraft in use to guarantee proper responses in all range of negative influences.

Full information on its work is necessary for complete control of a condition of the engine:

1. Reliable detection of a fault cause providing decision-making on a technical condition of gas turbines;
2. Reliable diagnosis and localization of faults and negative influences are necessary for definition of technical condition of gas turbines for the purpose of providing a reconfiguration and functioning of its subsystems [1, 2].

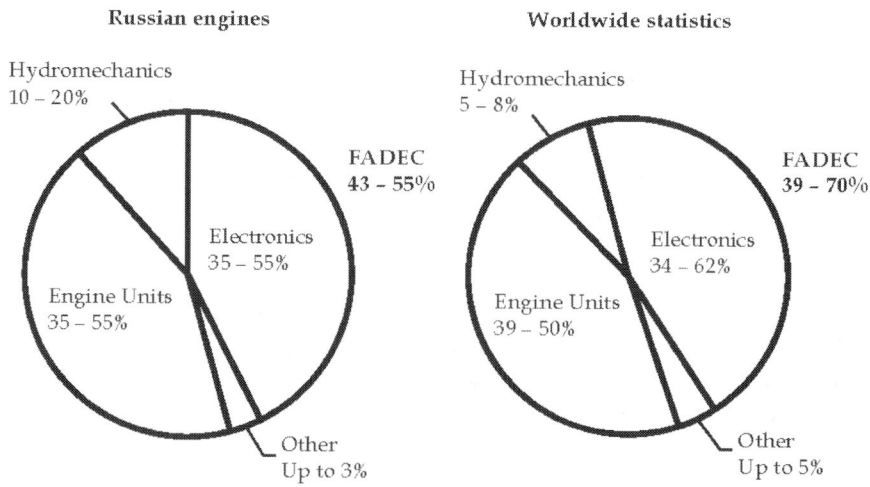

Figure 1. Faults percentage for engines and FADEC

The hardware for condition monitoring of measurement channels in many cases allows to detect only catastrophic (breakage or short circuit) faults, i.e. their stochastic properties on time of the process observed in one object and on a set of objects are not distinguishable [3]. The criteria of warning messages on faults appearance are based mainly on determined logic operations and distinguish between only two conditions: "operational" (fully operational) or "fault". In this chapter, hierarchical fuzzy Markov models for quantitative estimation of system safety of gas turbines taking into account the monitoring of cause-effect relations are considered. Transition from two-valued to fuzzy logic for estimation of degradation indexes and the analysis of fault developments for the gas turbine and its FADEC is considered for this purpose.

HIERARCHICAL MODEL OF FAULTS DEVELOPMENT PROCESSES IN GAS TURBINES

Complex diagnostics of the power plant is proposed to be carried out on elements and units, using the hierarchy analysis method [4, 5]. First, decomposition into independent subsystems of various hierarchy levels is carried out on structural features. Similarly, the power plant and its systems are represented in the form of hierarchy of elements and blocks.

This approach enables cause-effect relationships to be identified on the hierarchy structure of a system.

In Figure 2, the hierarchical structure of states of the power-plant is shown. The power plant is represented in the form of a hierarchical structure as the complex system consisting of subsystems and elements (units) with built-in test/monitoring functions, according to the distributed architecture. For this purpose, the power plant decomposition might be performed into independent subsystems with various levels of hierarchy on structural and functional features in the following way:

- Control and monitoring system (FADEC);
- Hydro mechanical system (actuators);
- Fuel system;
- Start-up system;

- Lubricant oil system;
- Drainage system, etc.

The hierarchy analysis allows to utilize the state model on the basis of faults development which enables the system state to be estimated at each level of the hierarchy.

The mathematical model of states is represented as

$$S= <G,F,L,R>,$$

where S is state vector,

G is hierarchy of system faults,

F is quantitative estimate of faults,

L is set of fault influence indexes,

R is mutual influence system of faults.

The depth of hierarchy G is referred to as h, and $h = 0$ for the root element of G.

For G the following conditions are satisfied:

1. There is splitting of G into subsets of h_k, $k = 1 \dots n$.
2. From $x \in L_k$ follows that $x-\subset h_{k+1}$, $k=1,\dots,n-1$.
3. From $x \in L_k$ follows that $x+\subset h_{k-1}$, $k=2,\dots,n$.

For every $x \in G$ there is a weight function such as:

$$\omega_x : x^- \to [0,1]; \text{where} \sum_{y \in x} \omega_x(y) = 1.$$

The sets of h_i are the hierarchy levels, and function ω_x is a function of fault priority of one level concerning the state of the power-plant x. Notice that if $x-\subset h_{k+1}$ (for some level of h_w), then ω_x can be defined for all h_k, if it equals to zero for all faults in h_{k+1} which do not belong to $x-$.

The hierarchical FADEC model integrates:

- functional structure (block diagram);
- physical structure;

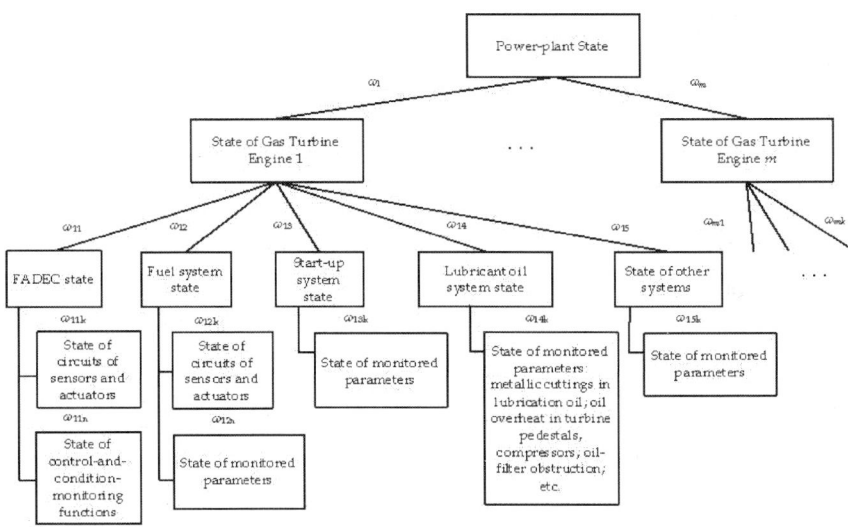

Figure 2. Hierarchical state structure of power plant

Table 1. Fragment of hierarchical classification of faults for FADEC

Fault levels	Faults		State	Fault handling priority
Level 1	FADEC fault Fault of lane B		Fault	
Level s 2-4		Fault of control function (B)	Degradatio n	Immediat e
		Fault of control loop (B)		

Level	Sensors	Actuators	
Level s 5-6	Integrated fault of measurement for alternative control law	Fault of actuator control circuit (B)	
Level 7	Fault of lane A		
Level 8		Fault of control function (A)	
	Fault of control loop	Fault of control loop (A)	
Level 9	Integrated fault of measurement (fault of same measured parameters in channels A)	Fault of actuator control circuit (A) (fault of communication lines of sensors or fault of FADEC hardware)	Short-term
Level 10	Fault of measurement in channel (break or short circuit)		Long-term
	Sensors	Actuators	

- tree states (state structure) of elements and units;
- tree of failures influence indexes.

On the hierarchical model, the system of faults interference R with logical operations of a disjunction and a conjunction is applied. Such a system of faults interference allows to analyze the state of all power-plant, both from the bottom up to the top, and from the top down to the bottom and to carry out deeper analysis on various levels of decomposition of the control system using an intermediate state: degradation.

The degradation is understood as "package/complex of degradationary changes of the system" and the degradationary change is "a separately considered irreversible change of a structure of the system, worsening its properties, changing the parameters and characteristics".

Define the main faults of FADEC, the priorities of their elimination and the state they belong to. InTable 1, an exemplary of a fragment of the hierarchical classification of faults for sensors and actuators of FADEC is presented.

Fault levels 1 through 6 demand immediate handlings and correspond to the "catastrophic" and "critical" states by FAA classification (Federal Aviation Administration, U.S. Department of Transportation), given in [6]. The emerging of such states requires immediate landing of the aircraft. Fault levels 7 through 9 are classified as a "marginal" state and demand operative handling after landing. In this case, it is possible to continue the flight, but postflight repair on the ground is required. Faults at the 10th level demand their handling in long-term prospect.

The degradation process for FADEC starts at the 10th level of hierarchy. From the 4th level of hierarchy, the system starts to approach the system crash that can be regarded as «a critical situation».

Note that development of such faults in certain cases can be detected in advance by estimating the states of elements not only at the level of "0-1" (fully operational, operational/working, fault), but also by considering their gradual degradation.

The state of an element or a system is proposed to be represented in the form of three parameters { operational, degradation, fault }, see Tab. 2.

In the operational state $S = 0$, while during fault $S = 1$. The degradation degree range from "0" to "1". Thus, the extreme values "0" and "1" are defined according to the determined logic, which is realized in the conventional FADEC (according to the design specifications for the system). The introduction of this intermediate state of "degradation" expands the informativity of the conventional condition monitoring algorithms.

Table 2. Fuzzy representation of state

Operational	Degradation	Fault
$S = 0$	$0 < S < 1$	$S = 1$

Based on the faults analysis and the hierarchy of states of the system at each level, the degree of degradation of each item or sub-unit is determined (Figure 3). Fault states are classified via degradation degree as "Negligible", "Marginal", "Critical" and "Catastrophic" [6]. The estimation of the degradation degree is defined on the membership function S which takes values in the range of $S \in [0,1]$. If the degradation degree is closer to "1", the distance to a critical situation will be closer. If the analysis of a system showed that the state vector is { 0,1 0,6 0,3 }, it is possible to ensure that there is a "distance" before complete fault (a critical situation). As soon as the system state will worsen with the appearance of new faults and will give the following state vector { 0 0,3 0,7 }, then there will be a distance of 0,3 to a system crash. Thus the most informative indicator will be a tendency of faults appearance (trend), not the existence of degradation itself. Visual trend analysis provides an estimate of time before the critical situations develop and, thus, for early planning of the crew actions [7, 8].

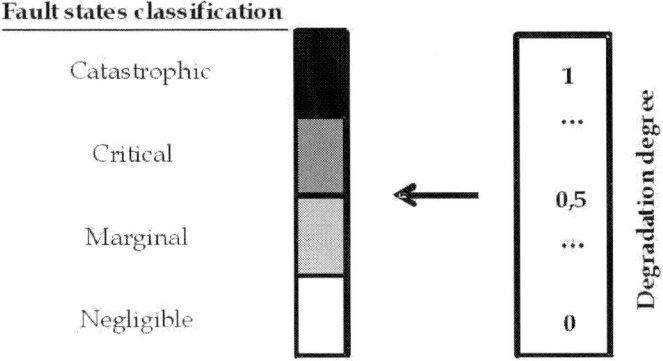

Figure 3. Estimation of "degradation" state

Consider an example of correspondence of degradation degree and the operational state. At the degradation degree of 0,25, the system is capable to carry out 75% of demanded functions (50% at 0,5 degradation, 25% at 0,75 and 0% at 1, which is the unavailable state). Such scale allows to define a "threshold" state, below which further operation is not allowed for safety reasons. Using the degradation degree, it is also possible to estimate the distance to a critical situation and the speed of approximation to it (Figure 4).

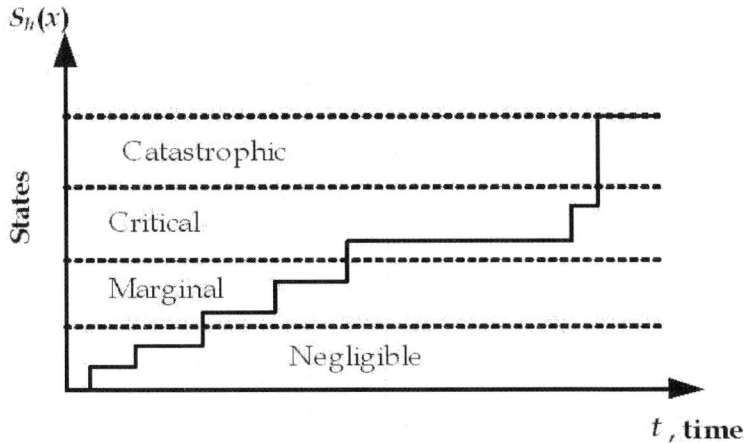

Figure 4. Trend of state dynamics during flight

Thus, the hierarchical model of fault developments allows to decompose the power-plant on hierarchy levels for obtaining quantitative estimates of the degradation state and gradual faults. The hierarchy analysis allows to utilize the state model and to estimate the system state at each level of hierarchy. The state is represented in the form of a vector with parameters { operational, degradation, fault }. Depending on the degradation degree it is possible to make an estimation of operability of object and system safety.

FUZZY TECHNIQUE OF DETERMINATION OF STATE PARAMETERS OF GAS TURBINE AND ITS SYSTEMS

The built-in monitoring system (BMS) is a subsystem of monitoring, diagnosis and classification of faults of the gas turbine and its systems. The fault existence corresponds to a logic state of "1", the absence does to a logic state of "0". Such state classification doesn't allow to establish a "prefault" state, to trace faults' development, and to define degradation of the system and its elements. For more detailed analysis, the estimation of the intermediate state of degradation is proposed. For this purpose, the use of fuzzy logic is considered. Signals from sensors, and also logic state parameters from BMS will transform to linguistic variables during fuzzification to a determined value arrives to the input of the fuzzifier. Let x is the state parameter of an element (for example, the sensor). It is necessary to define fuzzy spaces of input and output variables, and also terms for FADEC sensors. All signals from sensors and actuators will transform into linguistic variables by fuzzification.

Consider an example of fuzzy representation of a state of the two-channel sensor of rotational speed of a high pressure turbocompressor rotor. For monitoring of operational condition of communication lines of the FADEC sensor, a linguistic variable is introduced in the following way:

$$\Omega = <xn, B=(xn), U, G, M>,$$

where Ωn is the sensor state,

xn is the number of events when n is beyond the allowed limit band;

B is { operational, fault };
U is $[0,4]$
G is the syntactic rule generating terms of set B,
M is the semantic rule, which to each linguistic value x associate with its sense of $M(xn)$, and $M(xn)$ designates a fuzzy subset of the carrier U.

Say that the sensor is considered failed after the fourth appearance of the shaft speed measurement beyond the allowed boundary, therefore the membership function is formed as shown in Figure 5. At a single appearance out of limit ($xn = 1$), membership function B_1 takes the value 0,7, and $B_2 = 0,3$. The degradation degree takes the value of membership function B_2. If the repeated breaking the limit doesn't prove to be true during the set period of time, the monitoring algorithm cancels the measurement: $B_1 = 1$, $B_2 = 0$.

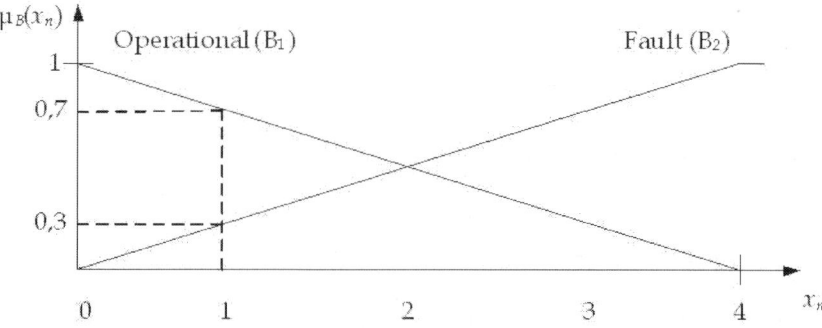

Figure 5. Membership functions of "sensor state"

In Figure 5 two membership functions are shown: state B1 corresponds to the function $\mu B_1 (xn)$, B_2 is described by the function $\mu B_2 (xn)$.

The way of creating fuzzy rules is presented in Figure 6. This rule base is represented by the table, which is filled in with fuzzy rules as follows [9]:

$$R(1):IF(xn=A1ANDxn=B1)THENy=T1$$

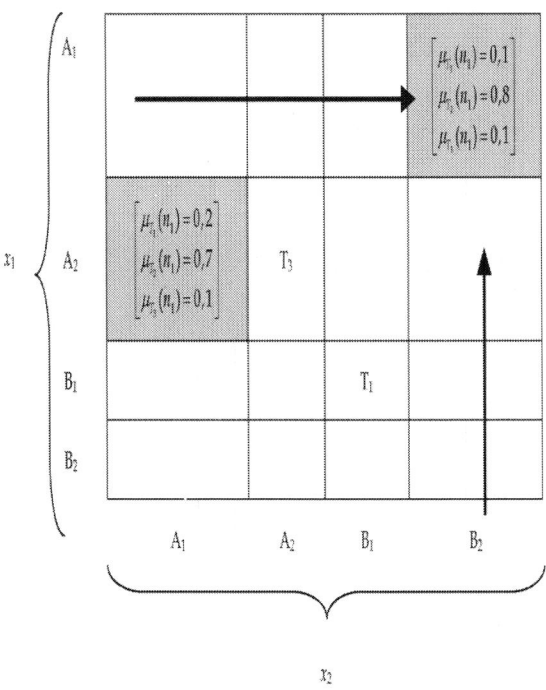

Figure 6. Example of fuzzy rule base

The values μT_1, μT_2, μT_3 are set in the cell at the row "Operational" (A_1) and the column "Fault" (B_2).

Consider a fragment of the rule base for estimates of the sensor state of the low pressure shaft speed. Formulate the first rule: if the 1st coil (n_{11}) of the sensor is operational (A_1) and the 2nd coil (n_{12}) of the sensor is operational (A_1), then the sensor is in the operational condition with the following membership functions:

[μoperational(n1)=1;   μdegradation(n1)=0;   μfault(n1)=0].

Write down this rule as follows:

$R^{(1)}$:IF(n_{11}=A_1ANDn$_{12}$=A_1)THENy=$T_1 \Rightarrow [\mu T_1(n_1)=1$; $\mu T_2(n_1)=0$;
$\mu T_3(n_1)=0$].

Other rules are created in the similar way.

Given a greater number of possible conditions (for example, greater number of the duplicated coils of the sensor), one can develop a discrete-ordered scale of state parameters (Figure 7).

For further analysis of the system, enter the faults influence indexes at each level of hierarchy, using a method of pairwise comparison as it is carried out in the hierarchy analysis method.

Quantitative judgements on the importance of faults are performed for each pair of faults (F_i, F_j) and these are represented by matrix A of the $n \times n$ size.

$$A=(aij), (i,j= 1,2,3).$$

where aij is the relative importance of fault Fi in regard to Fj. The value Aij defines the importance (respective values) Fi of faults in comparison with Fj.

Elements aij are defined by the following rules:

1. If aij =α, $aji = 1/\alpha$, $\alpha \neq 0$.
2. If   fault Fi has   identical   relative   importance   with Fj, then aij =1, aji=1, in particular aii=1 for all i.

Thus, a back-symmetric matrix A is obtained:

$$A = \begin{bmatrix} 1 & a_{12} & \cdots & a_{1n} \\ 1/a_{12} & 1 & \cdots & a_{2n} \\ \cdots & \cdots & \cdots & \cdots \\ 1/a_{1n} & 1/a_{2n} & \cdots & 1 \end{bmatrix}.$$

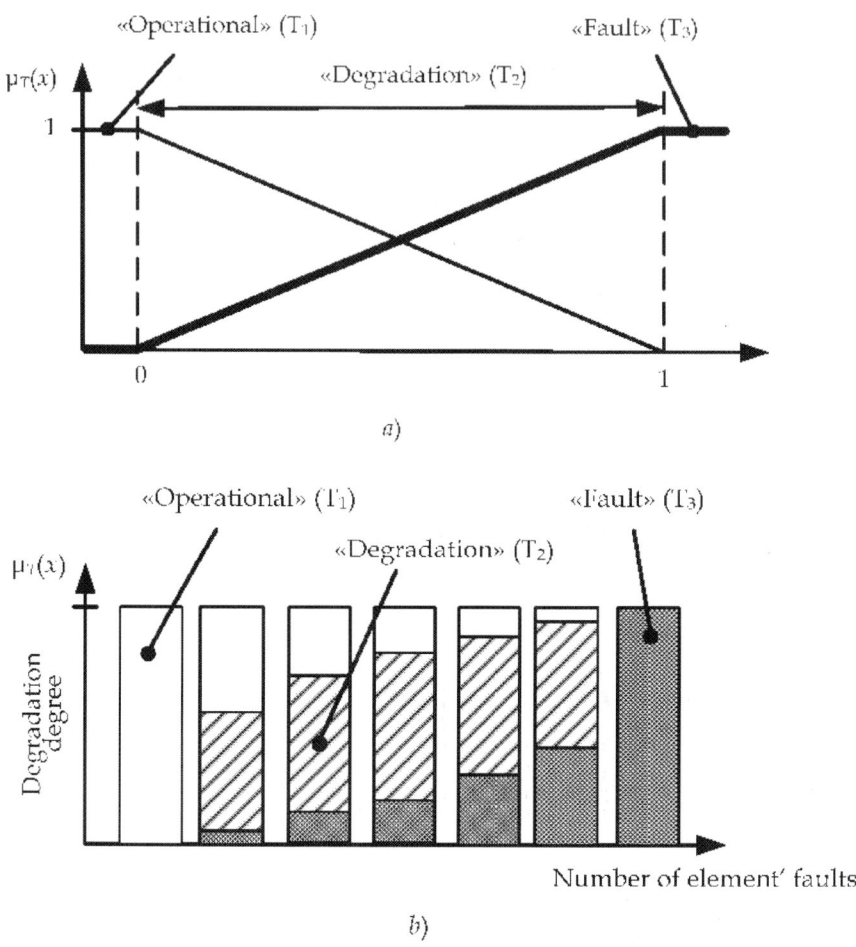

Figure 7. Continuous (a) and discrete (b) scale of degradation

After the representation of quantitative judgements about the fault pairs (Fi, Fj) in a numerical expression with the numbers aij, the problem is reduced to that n possible faults $F_1, F_2, ..., Fn$ will receive a corresponding set of numerical weights $\omega_1, \omega_2, ..., \omega n$, which would reflect the fixed judgements about the condition of the gas turbine subsystem.

If the expert judgement is absolute at all comparisons for all i, j, k, then matrix A is called *consistent*.

If the diagonal of matrix A consists of units ($aij = 1$) and A is the consistent matrix, then at small changes in aij the greatest eigenvalue λ_{max} is close to n, and the other eigenvalue are close to zero.

Based on the matrix of pair comparison values of faults A, the vector of priorities for fault classification is obtained, along with vector ω satisfying the criterion:

$$A\omega = \lambda_{max}\omega,$$

where ω is the eigenvector of matrix A and λ_{max} is the maximum eigenvalue, which is close to the matrix order n.

As it is desirable to have the normalized solution, let's slightly change ω, considering $\alpha = \sum_{i=1}^{n} w_i$ and replacing ω with $(1/\alpha)\,\omega$. This provides uniqueness, and also that $\sum_{i=1}^{n} w_i = 1$.

$$A = \begin{array}{c} \\ F_1 \\ F_2 \\ \vdots \\ F_n \end{array} \begin{array}{cccc} F_1 & F_2 & \cdots & F_n \\ \left[\begin{array}{cccc} w_1/w_1 & w_1/w_2 & \cdots & w_1/w_n \\ w_2/w_1 & w_2/w_2 & \cdots & w_2/w_n \\ \vdots & \vdots & \vdots & \vdots \\ w_n/w_1 & w_n/w_2 & \cdots & w_n/w_n \end{array}\right] \end{array} \begin{bmatrix} w_1 \\ w_2 \\ \vdots \\ w_n \end{bmatrix} = \lambda_{max} \begin{bmatrix} w_1 \\ w_2 \\ \vdots \\ w_n \end{bmatrix}.$$

Note that small changes in a_{ij} cause small change in λ_{max}, then the deviation of the latter from n is a coordination measure. It allows estimating proximity of the obtained scale to the basic scale of relations. Hence, the coordination index

$$(\lambda_{max} - n)/(n-1)$$

is considered to be an indicator of "proximity to coordination". Generally, if this number is not greater than 0.1 then it is possible to be satisfied with the judgements about the faults importance.

At each level hi of the hierarchy for n elements of the gas turbine and its subsystems, the state vector {operational, degradation, fault} is determined, taking into account the influence coefficients of failures:

$$S_{hi}(x_n) = \mu_{hi}(x_n) \cdot \omega_i,$$

where $\mu_{hi}(x_n)$ is the membership function value of the element xn (degradation degree). To determine the element/unit state of the hierarchy at a higher level $Shi(xn)$ for the input states of low-level $S_{hi-1}(x_n)$ one stage of defuzzification is performed.

The output value

$$S_{hi}(x_n)$$

is presented in the form of the determined vector of state with parameters { operational, degradation, fault }.

The state estimation begins with the bottom level of hierarchy. The description of a state set obtained by means of fuzzification and deffuzification with the use of the logic operations of disjunction ∨(summing), and conjunction ∧ (multiplication), which are designated as follows:

logic operation "AND"

logic operation "OR"

In performing operation "AND" in the inference system of fuzzy logic, the terminal tops are summed in order to determine the general state at one level of hierarchy that is presented as follows:

$$X_\Sigma = x_1 \vee x_2 \vee \ldots \vee x_n$$

In performing operation "OR", the "worst" state vector is chosen, with the maximum parameters of degradation $\mu_{degradation}(x)$ or faults $\mu_{fault}(x)$. The selector of maximum chooses from the fault influence indexes the one that has the maximum value.

The use of the hierarchical representation allows a small amount of "short" fuzzy rules to adequately describe multidimensional dependencies between inputs and outputs.

FUZZY HIERARCHICAL MARKOV STATE MODELS

A promising approach to constructing intelligent systems of control, diagnosis and monitoring could be the stochastic modelling on the basis of Markov chains combined with the formalized hierarchy theory.

Within a fuzzy hierarchical model, consider fault development processes with the use of Markov chains. Such dynamic models allow to investigate the change of elements' states in time. Fault development can include not only single faults and their combinations, but also sequences (chains) of so-called "consecutive" faults [10, 11].

During FADEC analysis, classification, formalization and representation of processes of condition monitoring and fault diagnosis for the main subsystems of gas turbines (control, monitoring, fuel supply etc.) is carried out. These processes are represented in the form of Markov chains which allow to analyze the state dynamics of the power-plant.

The transition probability matrix of a Markov chain for modeling faults and their consequences, has a universal structure for all levels of system decomposition (Figure 8):

- system as a whole (power plant);
- constructon units;
- elements.

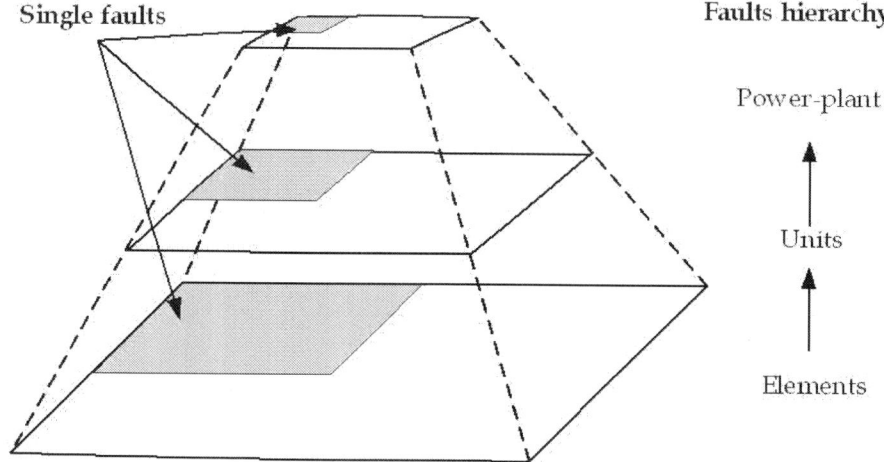

Figure 8. Hierarchical structure of Markov model of fault

The hierarchical Markov model is built in the generalized state space where physical parameters and binary fault flags are used for the estimation of a state vector of the element, unit and power-plant. The state vector includes three parameters { operational, degradation, fault } which allow to track the fault development and degradation process of the system. During FADEC diagnosis, the area of single faults is mostly considered. The proposed Markov model enables to present the system with multiple faults and their sequences. The top state level of a system reflects in the aggregated form the information on faults at the lower state levels.

The elements' state at the levels of the hierarchy depends on the previous values of state parameters of the elements, values of membership functions and fault influence indexes.

For the estimation of transition probabilities between the states of a Markov chain, it is required to calculate relative frequencies of events such as

$$Si \rightarrow Sj$$

for a given interval of time. In particular, at the top level, the number of events during one flight (Figure 9) can be of interest.

$P_{ij} = \mathrm{Prob}\{S_i \rightarrow S_j\}$	1	2	3
1. Operational state	P_{11}	P_{12}	P_{13}
2. Degradation	P_{21}	P_{22}	P_{23}
3. Fault leading to engine stop	P_{31}	P_{32}	P_{33}

Engine restart in flight

Figure 9. Transition probability matrix of power-plant during one flight

The most important events during flight are the emergency turning off of the engine stop (shutdown) and the possibility of its restart. For the probability estimation of such events, it is required to use statistics on all park of the same type engines. For realization of such estimation methods, it is required that flights information was available on each plane and power-plant. Such information should be gathered and stored in a uniform format and should be available for processing. Modern information technologies open possibilities for such research. To analyze fault development processes of one FADEC, it is possible to use results of the automated tests at the hardware-in-the-loop test bed with modeling of various faults and their combinations. In any case, to receive reliable statistical estimates one needs a representative sample of rather large amount of data.

In the analysis of the Markov model, the relation of the transition probability matrix with state of elements and subsystems at each level of hierarchy is considered. Therefore it is necessary to have the model of the system behavior in various states with various flight condition to guarantee system safety, reliable localization and accommodation of faults.

As the basic mathematical model of the controlled plant, the description in the state space is considered in the form of stochastic difference equations:

$$X(t+1)=AX(t)+BU(t)+F\xi(t), \tag{1}$$

where

$X \in R_s$ is the s-dimensioned state vector;

$U \in R_s$ is the s-dimensioned control vector; A, B and F are $(n \times n)$, $(n \times s)$ and $(n \times r)$ matrices accordingly;

$\xi \in R_s$ is the vector of independent random variables. Thus, the dynamic object described by this finite-difference equation, with input coordinate (control variable) U and output coordinate (state variable) X, in the closed scheme of the automatic control system is the controlled Markov process [12, 13].

The level of quantisation allows the Markov process to be converted into the Markov chain. Provided $\xi(t)$ is a stationary process, the Markov chain will be homogeneous. Such chain is described by the means of the stochastic transition probability matrix P with the dimensions $(m \times m)$, where m is the number of the chain states. Each element of the matrix Pij represents the probability of the system transition from the condition Xi into the condition Xj during the time interval ΔT:

$$Pij=Prob\{X(t)=Xi,X(t+1)=Xj\}, \quad \forall n \in N,$$

$$X_i \in \left[x_i - \frac{\Delta x}{2}; x_i + \frac{\Delta x}{2}\right]; \quad \sum_{j=1}^{m} P_{ij} = 1, i = \overline{1, m}.$$

$$(2)$$

Condition (2) means that the matrix P should be stochastic and define the full system of events. The sum of elements in each row of the stochastic matrix should equal 1.

The size of the matrix P is defined by the prior information on the order of the object model (1) and the number of the sampling intervals Δx and Δu. The transition probabilities are then estimated as relative frequencies of the corresponding discrete events.

The statistical estimation of the transition probabilities for the controlled Markov chain is performed as the calculation of the frequencies for the corresponding events during observation and the subsequent calculation of the elements of matrix P using the formula:

$$P_{ijk} = \frac{N_{ijk}}{\sum_{j=1}^{m} N_{ijk}}$$

$$(3)$$

where the numerator N_{ijk} is the number of the following events:

$$\{X(t_n) = X_i, X(t_{n+1}) = X_j, U(t_n) = U_k\}$$ and the denominator corresponds to the number of events such as $\{X(t_n) = X_i, U(t_n) = U_k\}$. Thus, for any combination of state Xi and control Uk, a full system of events will consist of the set of the state transitions Xj.

The normalisation of Equation (3) makes matrix P stochastic. As a result, the set of probabilities in each row Pij describes the full system of events for which the sum of probabilities is equal to unit:

$$\sum_{j=1}^{m} P_{ijk} = 1.$$

The estimation of a transition probability matrix of the Markov model consists of creation of multidimensional histograms which represent an estimate of joint distribution [14, 15].

The use of the hierarchical Markov model allows to "compress" information which has been recorded during one flight, and to present it in a more compact form. In this case, the possibility of analysis and forecast of dynamics of degradation degree (Figure 10) opens. It is possible to analyze the state dynamics of elements and functions at each level of hierarchy in time for decision-making support.

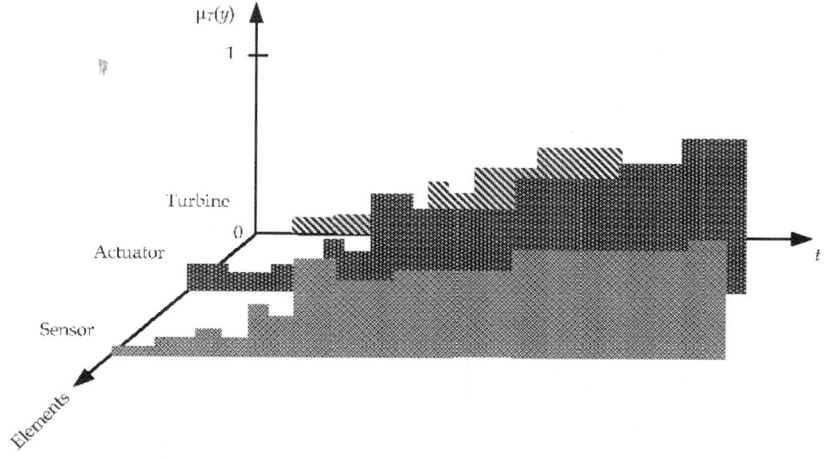

Figure 10. Dynamics of state parameters during flight

The analysis of fault information and state change can be carried out over flight data for the whole duration of maintenance and the whole "fleet" of engines and their systems (Figure 11). Such analysis will assist to increase efficiency for processes of experimental maintenance development and monitoring system support.

Given statistics on all park of engines within several years, it is possible to build empirical estimates of probabilities of the first and second type errors.

Thus, possibilities of application of hierarchical Markov models for the gas turbine and its FADEC for compact representation of information on flight and for the assessment of "sensitivity" of the monitoring system according to actual data are considered. The levels of hierarchy differ with the ways of introducing redundancy and realization of system safety with use of intellectual algorithms of control and diagnosis. Each higher level of hierarchy has greater "intelligence" and is designed independently in the assumption of ideal system stability of the lower level.

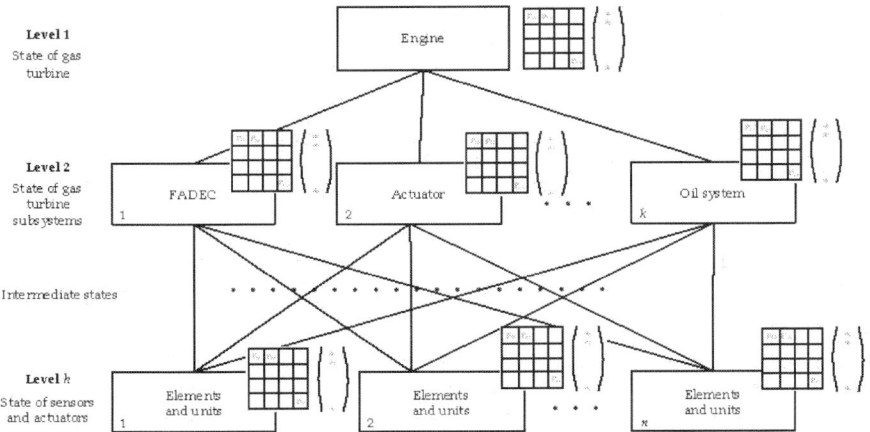

Figure 11. Hierarchical fuzzy Markov model of gas turbine states

Consider an example. In Figure 12, the FADEC state estimation with faults is presented on the basis of the degradation degree of the elements.

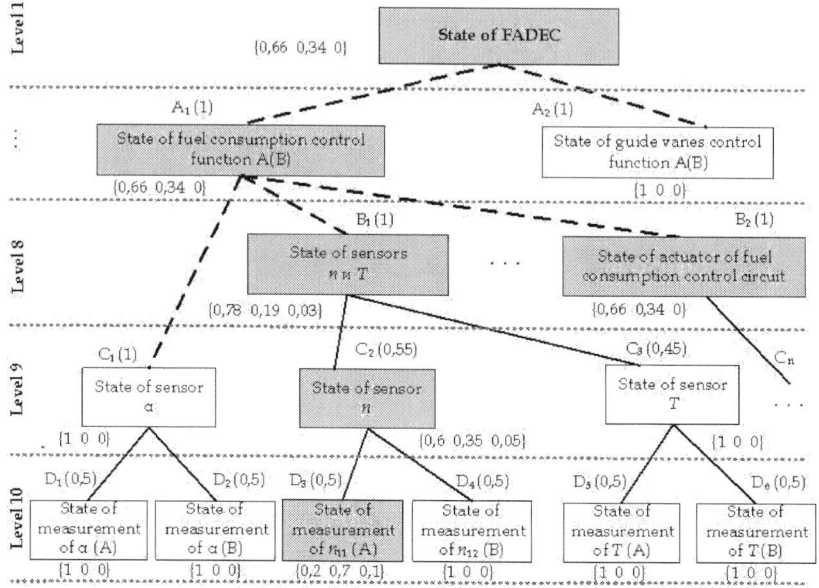

Figure 12. Example of hierarchical estimation of state parameters

At the 10th level the BMS detected a fault of measurement in the form of break of the first coil of parameter n_{11} (shaft speed sensor). On the basis of the fuzzy rule $R^{(2)}$, the parameters of measurement state of n_1 in the channel A are characterized by the following three values

$$\begin{bmatrix} \mu_{T_1}(n_{11}) = 0,2 \\ \mu_{T_2}(n_{11}) = 0,7 \\ \mu_{T_3}(n_{11}) = 0,1 \end{bmatrix}.$$

$$R^{(2)} : \mathbf{IF}(n_{11} = A_2 \mathbf{AND} n_{12} = A_1) \mathbf{THEN} y = T_2 \Rightarrow$$
$$[\mu_{T_1}(n_1) = 0,2; \quad \mu_{T_2}(n_1) = 0,7; \quad \mu_{T_3}(n_1) = 0,1].$$

The measurement state in the channel B is defined as

$$\begin{bmatrix} \mu_{T_1}(n_{12}) = 1 \\ \mu_{T_2}(n_{12}) = 0 \\ \mu_{T_3}(n_{12}) = 0 \end{bmatrix},$$

because no faults were detected. At the 9th level, the sensor state of the n is obtained using the multiplication of the vector of state parameters and faults influence indexes:

$$\begin{bmatrix} S(n_{11}, n_{12})_o \\ S(n_{11}, n_{12})_d \\ S(n_{11}, n_{12})_f \end{bmatrix} = \begin{bmatrix} \mu_{T_1}(n_1) = 0,2 \\ \mu_{T_2}(n_1) = 0,7 \\ \mu_{T_3}(n_1) = 0,1 \end{bmatrix} \begin{bmatrix} \mu_{T_1}(T_4) = 1 \\ \mu_{T_2}(T_4) = 0 \\ \mu_{T_3}(T_4) = 0 \end{bmatrix} \times [0,5;0,5] =$$

$$= \begin{bmatrix} (0,2 \times 0,5) + (1 \times 0,5) \\ (0,7 \times 0,5) + (0 \times 0,5) \\ (0,1 \times 0,5) + (0 \times 0,5) \end{bmatrix} = \begin{bmatrix} 0,60 \\ 0,35 \\ 0,05 \end{bmatrix}.$$

The state of an element of a higher level is calculated by multiplication of the current state to fault influence indexes of the fault elements. The state of both measurement channels of temperature T is "operational", therefore, the sensor T state equals

$$\begin{bmatrix} S(T)_o = 1 \\ S(T)_d = 0 \\ S(T)_f = 0 \end{bmatrix}$$

The sensor of fuel feed α is also good working.

At the 8th level the state of two sensors n and T after similar calculations becomes equal { 0,78; 0,19; 0,03 } that indicated the system degradation in the part of control of fuel consumption.

For the estimation of a state of the fuel consumption control function, the "OR" operation is also used. The state of the actuator of fuel consumption control circuit is characterized by the parameters { 0,66; 0,34; 0 }. The state of FADEC is characterized by the fault of fuel consumption control function or the state of the guide vanes control function. Using the operation "OR", the state of FADEC is detected as { 0,66; 0,34; 0 }. In this example, the whole system is considered to be operational, whereas partial degradation is observed, which is not influencing the system operability.

Thus, the technique of state parameters determination for FADEC and its systems on the basis of fuzzy logic and Markov chains is proposed. This technique can be used during flight or in maintenance on the ground.

At the present time a necessary condition for realization of intellectual algorithms is the complete development of the distributed intellectual control models focused on control optimization, forecasting and system safety [16, 17]. In Figure 13, the scheme of the distributed FADEC is shown.

Thus, in each sensor or actuator, it is necessary to have physically built-in control system (or function) to form and monitor the fault signals in the unit, communication lines, cooperating sensors, indication devices and systems [18]. The use of the built-in monitoring control systems working in real time allows to obtain a number of additional possibilities for improving control quality and system operational characteristics as followings:

- emergency states detection of the control object and system;
- fault detection of elements of the control object;
- state diagnosis and parametrical degradation of the object.

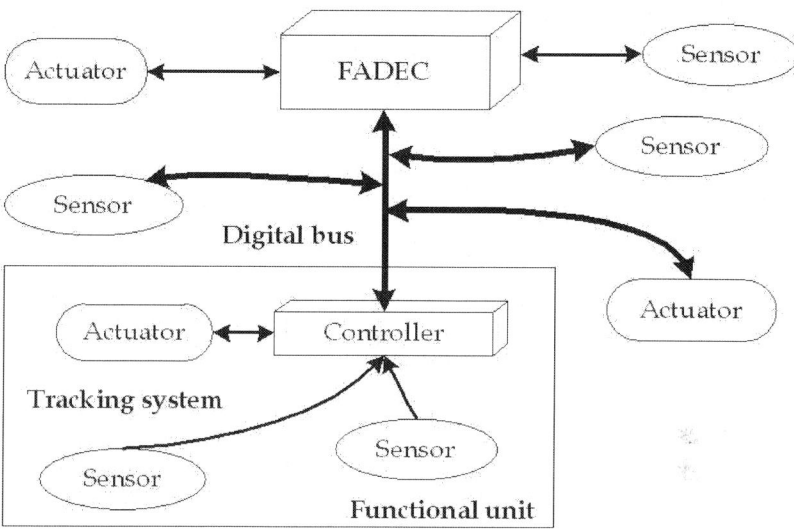

Figure 13. Distributed architecture of FADEC

CONCLUSION

In this chapter, the hierarchical fuzzy Markov modeling of fault developments processes has been proposed for the analysis of an airplane system safety. The hierarchical model integrates functional, physical structure of gas turbine and its FADEC elements and units, the tree of states, a tree of fault influence indexes. This model allows to decompose the power-plant for a quantitative estimates of degradation state and gradual faults. The analysis of hierarchies allows to utilize the state model on the basis of fault development processes which estimates the power-plant state at each level of hierarchy. Furthermore, the technique of determination of state parameters of the gas turbine and its systems on the basis of fuzzy logic is presented. The state of each element, unit and system is represented in the form of a vector with parameters { operational, degradation, faults }. The use of the proposed indicator "degradation degree" allows to obtain an objective quantitative estimate of the current state which can be used as, the "distance" to a critical situation and the reserve of time for decision-making in-flight. This indicator is defined on the basis of the discrete-ordered scale and fault influence indexes that allows to determine about 30 % of gradual faults in gas turbine and its systems at the stage of fault development. The examples of fuzzy rules on the basis of expert knowledge are given, whereas fuzzy logic is used for interpolation.

The application of hierarchical Markov models for the analysis of experimental data is also considered for control system development: as the compact representation of information of a system state change during flight, the estimation of transition probabilities.

Nomenclature

FADEC – Full Authority Digital Engine Control.
BMS – built-in monitoring system.

ω – weighting coefficient of fault.

S – state (condition) of gas turbines.

μ – membership function.

P – probability.

X – state variable.

U – control variable.

NaN. ACKNOWLEDGEMENTS

The work was supported by the grants from the Russian Foundation for Basic Research (RFFI) №12-08-31279, №12-08-97027 and the Ministry of Education and Science of the Russian Federation.

REFERENCES

1. Kulikov GG. Principles of design of digital control systems for aero engines. In Cherkasov BA. (ed.), Control and automatics of jet engines. Mashinostroyeniye, Moscow; 1988. p253-274.

2. Arkov VY, Kulikov GG, Breikin TV. Life cycle support for dynamic modelling of gas turbines. Prepr. 15th Triennial IFAC World Congress, Barcelona, Spain; 2002. p2135-2140.

3. Kuo BC. Automatic control systems. Prentice-Hall: Englewood Cliffs; 1995.

4. Saaty TL. Analytic Hierarchy Process: Planning, Priority Setting, Resource Allocation. McGraw-Hill, New York, London; 1980.

5. Saaty TL, Vargas LG. Models, Methods, Concepts & Applications of the Analytic Hierarchy Process. Kluwer Academic Publisher; 2001.

6. SAE ARP 4761. Guidelines and Methods for Conducting the Safety Assessment Process on Civil Airborne Systems And Equipment. Aerospace recommended practice; 1996.

7. Arkov V, Evans DC, Fleming PJ, et al. System identification strategies applied to aircraft gas turbine engines. Proc. 14th Triennal IFAC World Congress; 1999. p145-152.

8. Kulikov G, Breikin T, Arkov V and Fleming P. Real-time simulation of aviation engines for FADEC test-beds. Proc. Int. Gas Turbine Congress, Kobe, Japan; 1999. p949-952.

9. Zadeh LA. "Fuzzy algorithms," Information and Control 1968;12(2): 94-102.

10. Kulikov GG, Fleming PJ, Breikin TV, Arkov VY. Markov modelling of complex dynamic systems: identification, simulation, and condition monitoring with example of digital automatic control system of gas turbine engine. USATU, Ufa; 1998.

11. Breikin TV, Arkov VY, Kulikov GG. On stochastic system identification: Markov models approach. Proc 2nd Asian Control Conf ASCC'97; 1997. p775-778.

12. Breikin TV, Arkov VY, Kulikov GG. Application of Markov chains to identification of gas turbine engine dynamic models. International Journal of Systems Science 2006; 37(3) 197-205.

13. Kulikov G, Arkov V, Lyantsev O, et al. Dynamic modelling of gas turbines: identification, simulation, condition monitoring, and optimal control. Springer, London, New York; 2004.

14. Kulikov G, Arkov V, Abdulnagimov A. Markov modelling for energy efficient control of gas turbine power plant. Proc. IFAC Conf. on Control Methodologies and Technology for Energy Efficiency, CMTEE-2010, Faro, Portugal; 2010. http://www.ifac-papersonline.net/Detailed/42981.html

15. Arkov V, Kulikov G, Fatikov V, et al. Intelligent control and monitoring unit and its investigation using Markov modelling. Proc. IFAC Int. Conf. on Intelligent Control Systems and Signal Processing ICONS-2003, Faro, Portugal; 2003. p489-493

16. Vasilyev VI, Ilyasov BG, Valeyev SS. Intelligent Control Systems for gas Turbine engines. Proc. of the Second Scientific Technical Seminar on GT Engines, Turkey, Istanbul; 1996. p71-78.

17. Culley D, Thomas R and Saus J. Concepts for Distributed Engine Control, AIAA-2007-5709, 43rd AIAA/ASME/SAE/ASEE Joint Propulsion Conference and Exhibit, Cincinnati, Ohio, July 8-11, 2007.

18. Kulikov GG., Arkov VYu., Abdulnagimov AI. Hierarhical Fuzzy Markov Modelling for System Safety of Power-plant : Proc. of 4th International Symposium on Jet Propulsion and Power Engineering, September 10-12, 2012, Xi'an, China: Northwestern Polytechnical University Press; 2012. p. 589-594.

CITATION

G. G. Kulikov, V. Yu. Arkov and A.I. Abdulnagimov (2013). System Safety of Gas Turbines: Hierarchical Fuzzy Markov Modelling, Progress in Gas Turbine Performance, Dr. Ernesto Benini (Ed.), ISBN: 978-953-51-1166-5, InTech, DOI: 10.5772/54443.

CHAPTER 10

Semiconductor Light-Controlled Instrument Transducer With Direct Pwm Output for Automatic Control Systems

O. Malik, F.J. De la Hidalga-W. *

Instituto Nacional de Astrofísica, Óptica y Electrónica (INAOE)
Departamento de Electrónica Puebla, México

ABSTRACT

This work shows that the direct PWM output electric signal, with a duty cycle controlled by light intensity, can be obtained using a circuit containing a saw-tooth voltage generator connected in series with a dc voltage source and a metal (semitransparent gate) oxide semiconductor capacitor (MOS-C).

The internal PWM signal conversion occurs by the use of non-equilibrium physical processes in the semiconductor substrate of the MOS-C. The 10-20 V amplitude limited square PWM output signal is obtained by the amplification of the sensor signal with a standard 60 dB transimpedance amplifier. The amplified output signal presents positive and negative PWM waveforms that can be easily separated using diodes. The duty of the positive part is proportional to the light intensity, whereas the negative part is inversely proportional to the intensity. The frequency operating range of this proposed instrument varies from 1 Hz to a few kilohertz. The duty cycle of the PWM output signal varies from 2% to 98% when the incident light intensity varies in the microwatts range. These new transducers or sensors could be useful for automatic control, robotic applications, dimmer systems, feedback electronic systems, and non-contact optical position sensing for nulling and centering measurements.

INTRODUCTION

Among the several well-known optical sensors based on metal-semiconductor contacts or p-n junctions, metal-oxide-semiconductor capacitors (MOS-C) are known as the basic part of charge coupled devices (CCD) used in imaging technique [1].

The charging and discharging processes under irradiation in such devices are possible due to non-equilibrium processes in these semiconductor capacitors [2]. On the other hand, these processes are important for determining important physical parameters of the semiconductor substrate, such as the carrier concentration, generation and recombination times of minority carriers, and the characteristics of the semiconductor-oxide interface [[3], [4], [5] and [6]].

In this work we present new aspects of non-equilibrium processes in MOS capacitors with the aim of extending their optoelectronic applications as sensitive optical transducers with direct quasi-digital output in the form of a pulse-width-modulated (PWM) electrical signal. These sensors, with the new operating principle, allow for the simplification of electronic circuits necessary for automatic, robotic, and metrological applications.

We will also discuss the design and properties of a new optical position detector based on a bi-MOS structure that may be used as a sensitive null indicator with direct digital output in nulling and centering measurements.

OPERATING PRINCIPLE

Physical model

Sensors are based on non-equilibrium physical processes that become present in MOS capacitors which are initially biased by a constant voltage in the strong inversion operating mode. If a voltage pulse is added to this dc bias, an increase of the space charge region width (SCR) to a non-equilibrium value takes place. The excess minority carriers generated in the substrate lead to a reduction of the SCR width to its initial value, and the capacitor returns to its equilibrium mode. In absence of irradiation, the retention time depends on the generation rate of minority carriers in the substrate that is determined by the generation lifetime. The irradiation effectively decreases this time in comparison to that case under dark conditions. This fact is the basis of the new optical sensors. Let us qualitatively considerate the processes occurring in the MOS capacitor, with an oxide capacitance Cox, and initially biased by a dc voltage $U1$ in strong inversion. At a certain time, the additional triangular voltage with amplitude $U2$ is applied to the gate of the capacitor as shown in Fig.1.

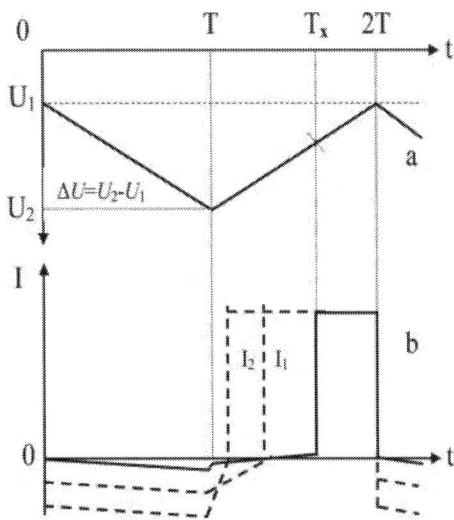

Figure 1. Schematic representation of the voltage (a) applied to the gate of the MOS capacitor, and the current (b) flowing through the capacitor: solid line- under dark conditions, dashed lines- under irradiance with $I_2 > I_1$.

At the first half period, the increase of the voltage from $U1$ to $U2$, with a rate $\Delta U/T$, leads to a time-dependent increase of the SCR width, from its stationary value $Winv$ to its maximum $W2$ at $t=T$ when a potential well is created. Both kinds of minority carriers, those generated thermally in the SCR, and those diffused from the neutral volume of the substrate, start filling this potential well, which decreases the SCR width.

If the flow of minority carriers is not too high, the potential well will not be filled completely during the first half period, and this process may also continue during the second half period until the voltage reaches Ux. At $t=Tx$, when the created potential well is filled, the MOS capacitor abruptly returns to its initial equilibrium state. At that time, the charge of the capacitor is $Cox(Ux-U1)$ and the displacement current presents the constant value $CoxdU/dt$. The irradiation increases the generation rate of minority carriers in the substrate, and the transition time becomes shorter than that obtained under dark conditions. This is shown in Figure 1b by dashed lines for two radiation intensities.

Mathematical model

Using absolute values, the triangular voltage bias shown in Figure 1 can be described by:

$$U(t) = U_1 + \frac{\Delta U}{T}t, \qquad (0 < t < T)$$

$$U(t) = U_2 - \frac{\Delta U}{T}(t - T), \qquad (T < t < 2T)$$

$$\tag{1}$$

The filling of the potential well with photo-generated minority carriers (the "charging" process) depends on the generation rate G. If the incident radiation has a low intensity, the charging of the MOS capacitor by the photo current occurs even when the functional voltage starts decreasing at the second half of the period.

When the potential well is filled, the reduction of the voltage till U_x at the time T_x leads to the returning of the capacitor to its initial equilibrium mode. At time T_x, the charge stored in the potential well during the photo generation is equal to the capacitor charge at the voltage (U_x-U1). This fact can be described as:

$$\frac{qGA}{\alpha}T_x = C_{ox}(U_x - U_1),$$

(2)

where q is the electron charge, α is the absorption coefficient of the semiconductor substrate at the wavelength of the incident radiation, and A is the gate area. The time T_x, as a function of the generation rate, can be found from (2) using the second equation of (1) for U_x:

$$T_x = \frac{2T}{1+K}, \quad K = \frac{qGAT}{\Delta U \alpha C_{ox}}$$

(3)

If the MOS capacitor is connected in series with a function generator (dc plus the triangular voltage), the output signal observed in a load resistor is a pulse with a duration of $2T-T_x$.

The duration of the output pulses and the duty D for the output signal (the ratio of the pulse duration to the period of the triangular voltage) are proportional to the irradiation intensity. This is because a higher intensity of the incident radiation provides a faster transition of the capacitor to its equilibrium mode. Thus, the output is a pulse width modulated electrical signal.

Taking into account (3), the equation for the output signal duty can be written as:

$$D = \frac{2T-T_x}{2T} = \frac{K}{1+K}.$$

(4)

Taking into account that $G=aPopt/Ah\nu$, where $Popt$ is the power of the incident optical radiation with a photon energy $h\nu$, the inverse dependence of the duty on the incident power can be written as:

$$\frac{1}{D} = 1 + \frac{qP_{opt}T}{AUC_{ox}h\nu}.$$

(5)

The duty does not depend on the area of the capacitor. Note that in the case of a triangular voltage with period $2T$, the maximum duty under irradiation can be only 0.5 when Tx is equal to the half period.

To obtain a higher range of duty due to the variation of the radiation intensity, a saw-tooth time dependent voltage must be applied to the gate. In this case, the sharp increment of the voltage from $U1$ to $U2$ during time t, which usually is 1-2% of T, creates the potential well, whereas the filling of this well by the photo generated carriers occurs during the slow decrement of the voltage from $U2$ to $U1$ during $T-\Delta t$. Then, using similar arguments as those used for the triangular voltage, the dependence of the duty on the incident radiation power is:

$$\frac{1}{D} = 1 + \frac{qP_{opt}(T-\Delta t)}{AUC_{ox}h\nu}.$$

(6)

In this case, the duty range may vary from 0.2 to 0.9 under irradiation.

There are some frequency limitations for both, the triangular and the saw-tooth voltages cases.

The low frequency limit is determined by the generation rate of minority carriers in the SCR as well as their diffusion from the neutral substrate. The low generation rate occurs at a long generation time of minority carriers. For example, if the generation time is ~ 0.5 ms, the transition time Tx will be close to 2 s, and the low frequency limit will be 0.5 Hz. Thus, the typical operating frequency for a non-equilibrium MOS-C under illumination corresponds to a few hertz. The high frequency limit is determined by the time response of the minority carriers with respect to the voltage variation and the reactance of the capacitor, and it lies in the range of a few kilohertz.

RESULTS

Fabrication and measurements of the MOS capacitors

To probe our conception, MOS capacitors were fabricated on high resistivity (2-4 kΩ-cm) n-type silicon substrates, with a 70 nm therma

lly grown oxide, and a titanium semitransparent gate. For measurements, the capacitor was connected in series with a function generator, with a dc offset and a load resistor as shown in Figure 2. The output was recorded using the digital oscilloscope TDS 3054C; a light emitting diode (LED), with emission at 0.9 nm, was used to illuminate the MOS capacitors.

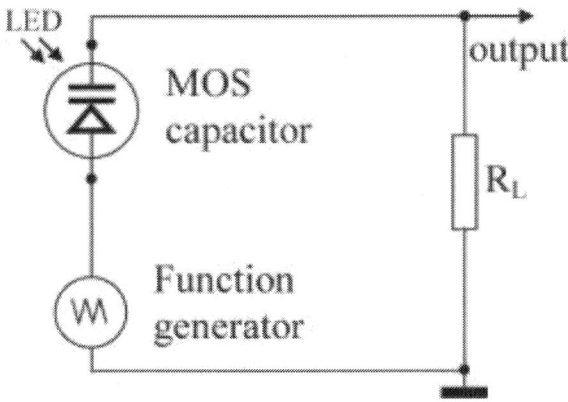

Figure 2. Circuit used for the measurement of the MOS-C optical sensor.

Experimental results and discussion

Figure 3 shows the current for the MOS capacitor under dark conditions after applying the dc and the triangular voltage. The oscillograms for the current flowing in the capacitor under illumination were recorded for an LED incident power varying in the 20 nW to 0.5 mW range.

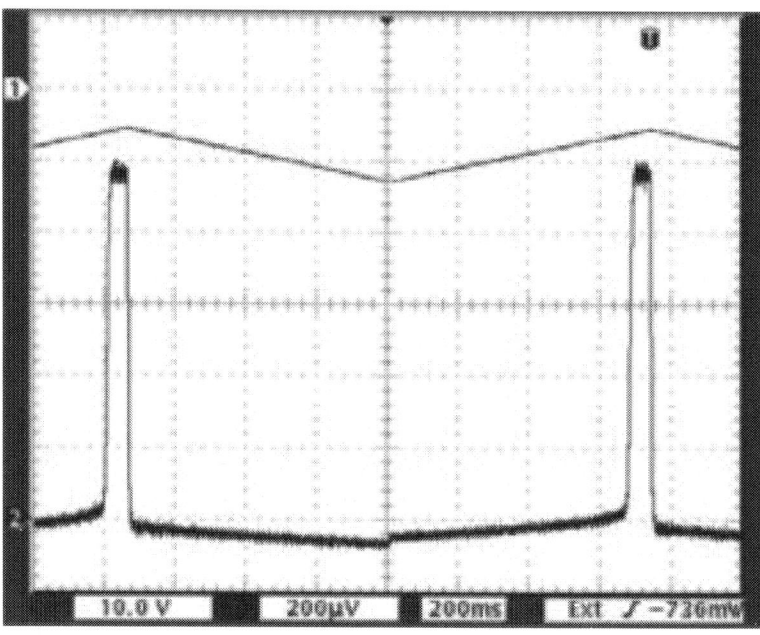

Figure 3. Oscillograms showing the applied dc and triangular voltages (above, scale 10 V/div) and current in dark (below, scale 200 μV) obtained with a 50 kΩ load resistor.

From Figure 4, one can see the difference in the transition time for the illuminated MOC-C, under different illumination conditions, when the capacitor was biased with the dc and triangular (60 Hz) voltages. The transition time decreases for an increasing illumination.

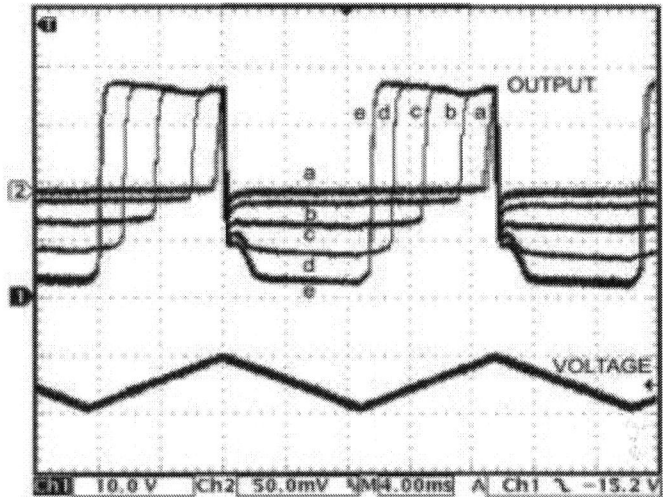

Figure 4. Oscillograms (4 ms/div) for the applied dc and triangular voltages (below, 10 V/div), and current (50 mV/div, curves a-e), obtained for the illuminated capacitor. The illumination increases from a to e.

Figure 5 shows the oscillograms for the current of the illuminated MOS capacitor when a saw-tooth voltage is applied to the gate.

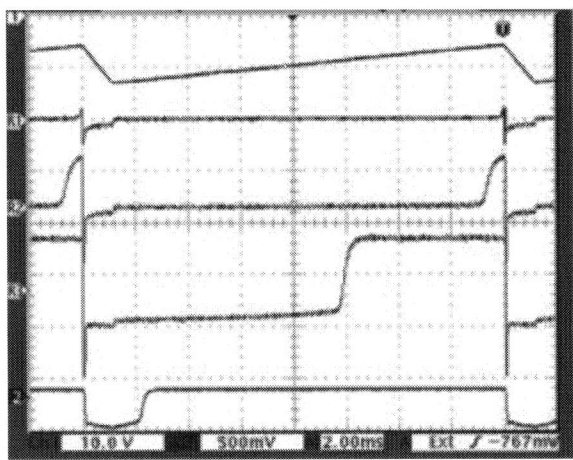

Figure 5. Oscillograms (2ms/div) for the applied voltage (above, 10 V/div) and current (500 mV/div) for dark (R1) and three irradiation conditions (R2, R3 and 2). The scale for the last current oscillogram was multiplied by ten.

To obtain a "clear" PWM output signal presenting sharp transitions at both edges of the pulse, the signal from the load resistor was amplified in 10^3 times, and limited by the amplifier to 10 V. The negative part of the signal due to the photocurrent was rejected using a diode.

Figure 6 shows the oscillograms for the amplified output for the dc and triangular voltages applied to the gate of the MOS capacitor.

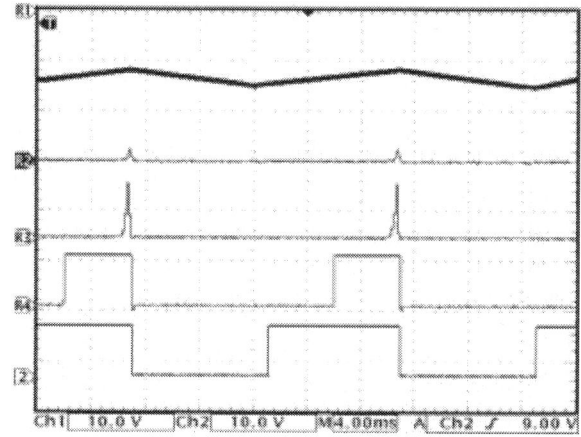

Figure 6. Oscillograms (4 ms/div) for the dc and triangular voltages applied to the gate (above curve, 10V/div), and amplified output signal(limited to 10 V) under different values of illumination from the LED (R2, R3, R4, and a).

It is evident the pulse-width-modulation nature of the output signal under different illumination levels. For the case of a triangular voltage, the maximum value for the duty is 0.5.

A higher range of variation for the duty, from 0.02 to 0.87, is obtained when a saw-tooth voltage is applied, as is shown in Figure 7.

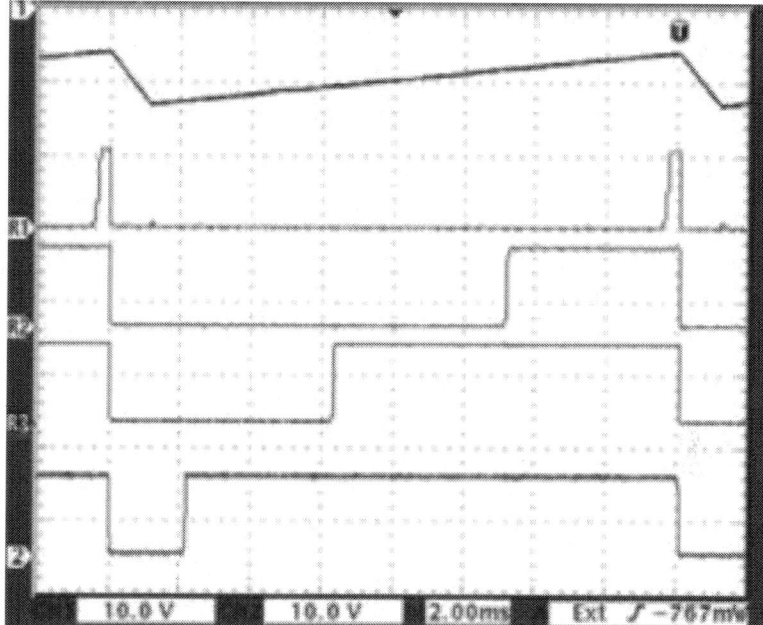

Figure 7. Oscillograms (2 ms/div) for the dc and saw-tooth voltages applied to the gate (above curve, 10V/div), and the amplified output signal (10 V) under different values of illumination (R1, R2, R3, and a).

From above, these non-equilibrium MOS capacitors can be used in applications such as digital sensitive optical sensors in automatic control and robotic circuits. Below, we present another new metrological application for these MOS capacitors.

Bi-MOS capacitor as optical digital position sensor for nulling and centering measurements.

The bicell optical detectors are well known for position sensing. They operate under the principle of having two photodiodes separated by a small gap. These elements are generally built onto a common substrate, thus their cathodes are shared. The anode or active area of each element is individually contacted. When a light spot is translated across the detector, its energy is distributed between both elements, and the difference in electrical contributions to each element defines the relative position of the light spot with respect to the center of the device. The detector provides the position information only over a linear distance of twice the spot diameter or until the edge of the spot has reached the detector gap. A linear transfer function can be obtained for a rectangular light spot because its linear movement is proportional to the percentage of its area that shifts between the elements. The sensitivity of silicon bicell position detectors is 2-3 V/mm. Such detectors are most effectively used as nulling and centering devices with resolutions of 0.1 micrometers or higher. Nevertheless, these devices present analog output, thus an analog-to-digital signal converter (ADC) is necessary for their application in modern circuits. Additionally, the characteristics of the ADC can affect the precision in position measurements. The use of non-equilibrium processes in MOS capacitors under illumination allows for designing new devices for position sensing.

Figure 8 shows schematically the construction of our new position sensitive optical sensor. Our conception of this new position sensor is based on two MOS capacitors (C_1 and C_2), fabricated on the silicon substrate separated by a distance longer than the diffusion length of minority carriers in the substrate. Part of this area is covered by a metallic opaque shield. The size of the light spot must be slightly longer than the width of the shied. If the spot is located at the centre of the distance between both capacitors, approximately the same generation rate will be found at the end of each capacitor. The electric field at the lateral SCR of the biased capacitor separates the photo-generated carriers, and an equal photocurrent will flow through each capacitor. Under such conditions, and as it was shown earlier, an output PWM signal with the same duty can be recorded. A small movement of the spot will produce a misbalance in both

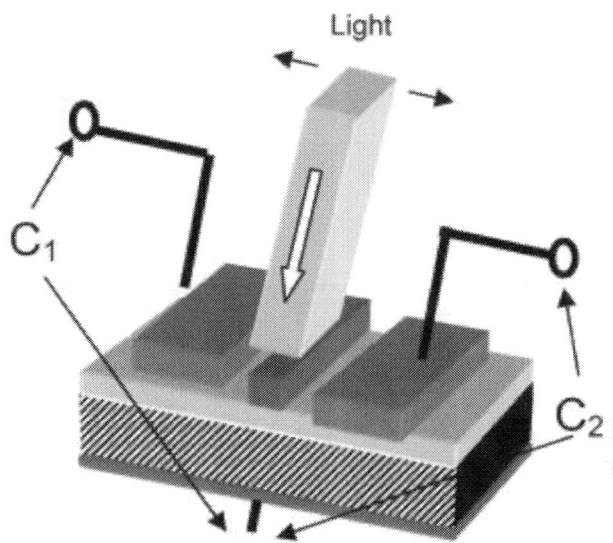

Figure 8. Schematic model of the new position sensitive optical sensor.

photocurrents and consequently a difference in duty for both output electric signals. If a differential amplifier is used to record these PWM signals, the information about the spot position may be obtained from the polarity and the duty of the resulting signal. Using (6), the mathematical modeling of the sensing characteristics of such position sensor has been provided for the silicon MOS capacitors, and a rectangular 2mm×1mm light spot with a power of 0.3 mW/cm² from a LED with emission wavelength at 930 nm. The initial light intensity provides two equivalent PWM output signals with a duty of 0.5. Taken into account a variation of the incident optical power during the spot movement, the differential PWM signal (that can be provided by a differential amplifier) was calculated as the dependence of the resulting duty (D) on the spot position. Figure 9 shows this dependence.

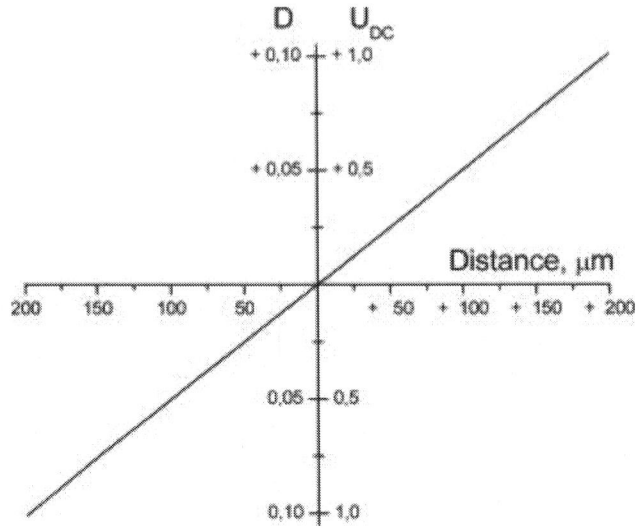

Figure 9. Calculated dependence on position of the light spot for the duty (D) and U_{DC}.

The transfer function is linear for the spot movement till 200 micrometers. At that limit the duty of the resulting PWM signal changes to 0.1. The polarity of the resulting PWM signal shows the direction of the movement. The use of a low-pass filter allows for transforming the 10 volts output PWM signal into a dc voltage (U_{DC}).

The slope of the dependence for U_{DC} allows for obtaining some conclusions about the sensitivity of this new position sensor. This sensitivity is two times higher than that obtained with standard bicell silicon analog detectors. We also conducted the experimental modeling of the reported sensors fabricated on a high-resistivity silicon substrate.

The PWM output signals from both capacitors were amplified with a two-channel transimpedance amplifier. The amplified signals were added with a differential amplifier and recorded using a digital oscilloscope. Each capacitor was connected in the circuit configuration shown in Figure 2.

The combination of the dc 10 V and the 10 V triangular voltages was applied to the gates of the capacitors. The oscillograms of the output signals with and without amplification, when the dc and triangular voltages are applied to the gate of MOS capacitors, are shown in Figure 10.

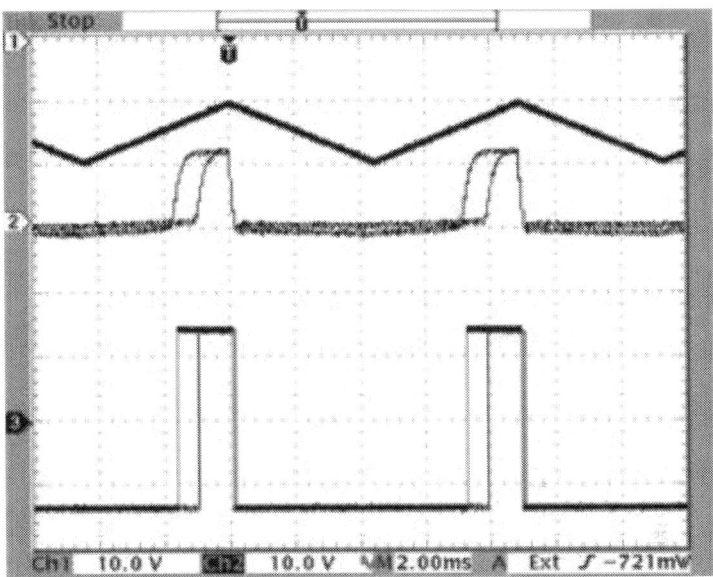

Figure 10. Oscillograms for the non-amplified (middle curves) and amplified (below curves) outputs obtained for the bi-MOS capacitor under non-uniform illumination conditions. The composition of the dc and triangular voltages (top curve) are applied to the capacitor gates.

Thus, the principal feature of our proposed sensor is the obtaining of a direct PWM output, with a duty that indicates the position of the light spot, whereas the polarity of the output signal indicates the direction of the spot movement.

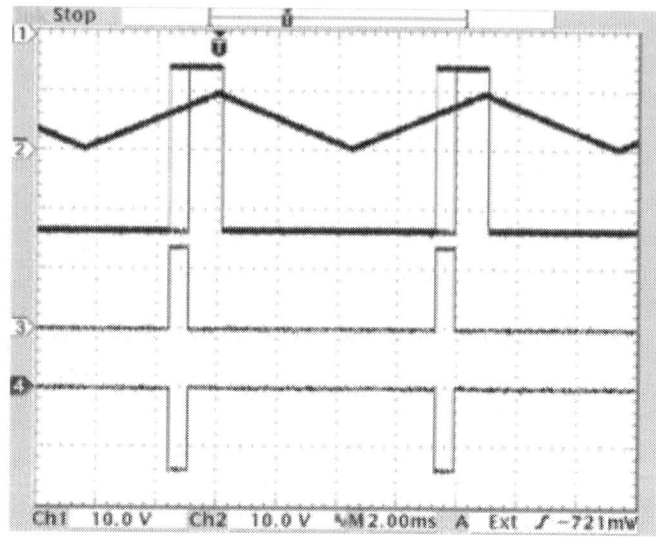

Figure 11. Added output signals obtained for the bi-MOS position sensor.

A detailed characterization of our new sensors for nulling and centering measurements will be presented in a separate publication.

CONCLUSIONS

In this work we showed that non-equilibrium processes in MOS capacitors can be used for designing new optical sensors with direct pulse-width-modulated output, with a duty proportional to the light intensity. The amplified digital output using a rectangular waveform allows for the direct connection to a microcontroller or other logic circuitry. Such sensors are useful for different applications in automatic control and robotics systems because of their fabrication simplicity, as well as the possibility to simplify electronic circuits. A new type of optical sensor for nulling and centering measurements is proposed using on-chip bi-MOS capacitors. Our previous mathematical and experimental modeling showed interesting features of these sensors as a possibility to fix the position of a light spot by observing the duty value of the differential output signal, and the moving direction of the spot through through the polarity of the differential output signal.

These sensors present a linear transfer function for a light spot movement of about 200 micrometers, and a two-time improved sensitivity in comparison with analog devices.

ACKNOWLEDGEMENTS

Authors thank the technicians of the Microelectronics Laboratory at the National Institute for Astrophysics, Optics and Electronics. This work was partially supported by CONACyT Mexico under grant 102397.

REFERENCES

1. S. Sze, "Physics of Semiconductor Devices", New York: John Wiley & Sons, 1981, pp. 407-430.
2. D. Schroder, "Semiconductor Material and Devices Characterization", New York: John Wiley & Sons, 1990, pp. 405-423.
3. K. Y. Cheong, Sima Dimitrijev, Ji Sheng Han, "Characterization of Non-Equilibrium Charge of MOS Capacitors on p-Type 4H SiC", Materials Science Forum, vol. 457-460, pp. 1365-1368, June 2004.
4. P. Peykov, J. Carrillo, and M. Aceves, "Triangularvoltage sweep C-V method for determination of generation lifetime and surface generation velocity", Solid-State Electronics, vol. 36, No. 1, pp. 99-102, January 1993.
5. M. Tapajna, L. Harmatha, "Determining the generation lifetime in MOS capacitor using linear sweep techniques", Solid-State Electronics, vol. 48, No.8, pp. 2339-2342, August 2004.
6. K. Ding, "Simple determination of the profile of bulk generation lifetime in semiconducto", Solid-State Electronics, vol. 46, No. 4,pp.601-602, April 2002.

CITATION

O. Malik, F.J. De la Hidalga-W., Semiconductor Light-Controlled Instrument Transducer with Direct PWM Output for Automatic Control Systems, Journal of Applied Research and Technology, Volume 11, Issue 1, February 2013, Pages 18-25, ISSN 1665-6423, http://dx.doi.org/10.1016/S1665-6423(13)71512-8.

Index